THE

Antient Bee-Mafter;

OR,

FULL AND PLAIN DIRECTIONS

FOR THE

Management of Bees to the greateft Advantage;

DISCLOSING

FURTHER IMPROVEMENTS

OF THE

HIVES, BOXES, AND OTHER INSTRUMENTS, TO
FACILITATE THE OPERATIONS;

Efpecially that of SEPARATING *Double and Treble Hives or Boxes,
with Certainty and Safety, without injuring the Bees*;

INTERSPERSED WITH

NEW BUT IMPORTANT OBSERVATIONS:

The WHOLE *ftudioufly adapted to* GENERAL USE;

WITH

AN APPROPRIATE METHOD FOR THE CURIOUS.

ALSO

BRIEF REMARKS ON SCHIRACH,

And other diftinguifhed Apiators on the Continent.

DEDUCED FROM A SERIES OF EXPERIMENTS DURING
THIRTY YEARS.

ILLUSTRATED WITH PLATES.

" Multum in Parvo."

BY JOHN KEYS, OF BEE-HALL, NEAR PEMBROKE.

LONDON:

PRINTED FOR G. G. AND J. ROBINSON,
PATERNOSTER-ROW.

1796.

PREFACE.

IMPROVEMENTS in the management of Bees, among farmers and cottagers, have been but little advanced, notwithftanding the *many* ingenious contrivances which have been offered to public notice; probably from being too *operofe* and *expenfive* for people of that defcription, to whom, in common, the management of bees is generally of trifling concern.

Whether my prefent attempt will fucceed better, time muft decide. My utmoft exertions have been directed to the remedy of the defect; which, it is hoped, I have greatly *advanced*, if not perfected as far as our unfavourable climate will admit.

Additional profits, in moft cafes, are unavoidably attended with fome increafe of

expence.

expence. But from the *impartial eftimate*
I have ftated (page 60), it will be afcer-
tained, that the management there propofed
far overbalances the extra expence, and
therefore will merit the adoption of Bee-
keepers, and, perhaps, the patronage of
Agricultural Societies.

By perfons of a higher clafs a trifle of
additional expenditure will not be regarded,
in comparifon of the convenience and fafety
with which the operations may be effected :
to fay nothing of the gratification of philofo-
phic curiofity, together with fuperior profit.

Near the clofe of the year 1780, I ven-
tured to publifh a work of this kind, according to the beft of the knowledge and experience I had then attained. Since that time,
having no avocation to withdraw my attention from my *favourite purfuit*, and en-
couraged by the favourable reception of that
work, I now, in the vale of life, fubmit this
treatife, as the *refult* of all my refearches ;
drawn from a much longer and more affi-
duous experience, and from a cooler judgment, ripened by numberlefs experiments,
which have led me to *new* obfervations and
                                        improve-

improvements, and to *differ* alfo not more from *myfelf* than from ALL OTHERS.

Inftead of a fecond edition of my former publication, a *new* book became neceffary, as moft part of my prefent management is on a different plan; and much tautology and fuperfluity of matter demanded curtailing, and a more judicious arrangement of the whole.

No article inculcated in thefe pages is advanced without its being warranted by my own experience, unbiaffed by any author, however otherwife refpectable. Where I am ftill dubious, it is fo expreffed.

Apiators may be affured that, to the beft of my knowledge, every information or hint that has been found of any *real* fervice, in any Writer of Note, Foreign or Domeftic*, is *comprifed in this volume.*

As my prefent defign is *wholly for practitioners*, the bulk and price is accommodated to the purpofe of becoming generally ufe-

---

* Butler, Mew, Geddy, Purchafe, Wolridge, Rufden, Warder, White, Thorley, Mills, Wildmans, Debraw, and Broomwich. Foreigners; Miraldy, Reaumur, Bonnet, Schirach, Needham, Norton, Seykers, and others of lefs note.

ful;

ful; and confequently precludes the Natural
Hiftory of Bees*, except in fome fmall degree, as far as neceffary to their management.

A few years fince, warm difputes arofe
between different naturalifts and apiarian
focieties on the continent, relating to the
generation of bees, and the formation of
artificial fwarms, in confequence of fome
new and *wonderful* principles advanced by
a Mr. Schirach (fecretary of an apiarian
fociety), in his treatife entituled " *Hiftoire
Naturelle de la Regne des Abeilles*," &c. tranf-
lated into French by J. Blaffiere, Hague,
printed 1771†.

Counter-experiments were made by Needham, Rheim, and others‡, with refults of
an oppofite nature.

It being incompatible with the defign of
this work to enter into details, or a formal

---

* See a judicious book with that title, being a compi-
lation from the French, publifhed by Knapton 1744.

† This book I had the honour of having prefented
me by the late *Earl of Marchmont*.

‡ Bruffels Memoirs, vol. ii. 1780, prefented me by
Thomas Dilks, Efq. to whom I render my thankful
acknowledgments.

refu-

refutation of Schirach's doctrine, I shall only briefly declare, that at first I was strongly prejudiced in its *favour*, and urged thereby to purfue a feries of experiments according to his directions, with the moft fcrupulous exactnefs and care, for eight years, but without a SINGLE RESULT in confirmation of his fcheme. I diverfified the experiments, and alfo invented a more fuitable apparatus to perform them, yet ftill met with the fame lamentable difappointment. In this purfuit many bees and many ftocks were unavoidably ruined, befides an accumulation of vexation and trouble. But my anxiety to acquire fo defirable an end urged me beyond the bounds of prudence. I hope vanity will not be imputed to the declaration and inference, that if one of long experience in the handling of bees, and having every conveniency, yet could not, in fo many trials, fucceed, it is more than probable that others, with only common knowledge, and deftitute of a proper apparatus, will not be more fuccefsful, and confequently that Schirach's method cannot prove of public utility.

The

The moſt likely means to *eſtabliſh* the BEE ART, I believe, will not be accompliſhed without the PATRONAGE of *Agricultural Societies*; namely, the encouraging a proper perſon or perſons to exhibit in the *bee ſeaſon* the moſt *approved method* through the chief market towns of the kingdom. The perſon muſt be capable of explaining the proceſſes, and have with him the neceſſary inſtruments.

I would alſo *recommend* this employment to any ingenious young man, properly qualified, and provided with the apparatus, as a practice that moſt likely would turn to much advantage; taking care *not* to introduce tricks and fancies, as ſome have done, to the *deſtruction of multitudes of bees*, inſtead of exhibiting any real improvements.

Or, if *ſuch* perſons, reſident in villages, would qualify themſelves, they might, even in their limited ſtations, exerciſe the art to their own benefit and that of the neighbourhood, by performing it at a *ſtated rate.*

Rural Curates might conſiderably augment their too frequently *niggardly* ſtipends,

by

by the cultivation of bees, and act at the same time confonantly with their clerical profeffion, as it is an innocent amufement, both *healthy* and *profitable.*

Farmers and others who keep numerous ftocks of bees cannot be fuppofed to attend fo minutely to every particular as thofe who have not many, and have more leifure; yet in this, as well as in all the other articles of hufbandry, the greater the care beftowed, the greater the return that ufually follows. Befides, moft of the operations on bees are to be performed in the evening, or early in the morning, and therefore will not interfere with more important bufinefs.

To avoid repetitions, the *manner of performing the operations* muft be feverally learnt from the *fecond part,* to the particulars of which the Index will direct.

The firft part contains the principles; the fecond the manual, or operative part.

In general, I have ufed the word *hive* indifcriminately for *hive* or *box,* as applicable to either; except when it is otherwife expreffed, or is felf-evident.

The

The writer submits thefe pages to the candour of the learned, under a confciouf-nefs of his own inabilities for the tafk ; but if it affords *ufeful* improvements in the art, he hopes it may be pleaded in excufe of his prefumption.

JOHN KEYS.

Bee-Hall, near Pembroke,
1796.

EXPLA-

EXPLANATION OF TERMS.

ADAPTER, a Board to fet Glaffes on.

Apiary, the Place or Spot where Bees are kept.

Apiator, the Perfon who manages the Bees.

Bee-Herd, the Perfon who watches the rifing of Swarms.

Cafts, fecond, third, &c. Swarms.

Deprivation, the Separation, or Taking of the Hives of Honey.

Divider, the Brafs Plate ufed in feparating Hives.

Duplet, the Hive fet over or under another.

To Duplicate, the act of performing this.

Fume-Box, the Box kept for the Purpofe of Fuming.

Fumigation,
Fuming, } the Operation of ftupefying.

Hackel, or *Coppet*, &c. the Straw Covering fet over a Hive to fhelter it.

Hive, or *Skip*, &c. *That* wherein the Bees dwell, and make their Combs, whether made of Straw or other Materials.

Nadir, the Hive which is fet under another.

Non-fwarmer, a Stock which has not fwarmed.

Stock, a Hive of Bees that has ftood, or is to ftand, the Winter.

Storifying,

Storifying, the ranging Hives over or under each other.

To Storify, to perform this.

Super-hive, to set one Hive above another.

Superior Hive, the uppermost of a storified Set.

Swarm, a great Body of Bees, which quit the Hive together, and fix in some Tree, Bush, &c.

To Treble or *Triple,* to add a third Hive to a Stock that has two before.

Trebled, a Stock that has three Hives.

Triplet, ditto.

To Triplicate, the Act of triplifying.

CONTENTS.

CONTENTS.

PART I.

PART II.

THE OPERATIONS.

Antient Bee-Master's Farewell.

PART I.

CHAP. I.

OF THE QUEEN, DRONES, AND WORKERS.

ON the *fingle female bee*, ftyled QUEEN, depend the increafe, profperity, and permanency of a ftock. No fwarm can poffibly be eftablifhed, unlefs accompanied by a princefs; although the bees become ever fo numerous, or eager to fwarm. If by any mifchance the queen is killed, the bees, foon fenfible of her lofs, quit the hive to affociate with their next neighbours, transferring their treafure with them.

B   The

The QUEEN (pl. I. fig. 6.) being then of fuch confequence, it is neceffary that the apiator fhould be able to diftinguifh her at fight. Obferve, therefore, that fhe is longer and more flender than the drones, or the workers; her hinder parts tapering to a point: her belly and legs are alfo yellower; and the upper part of her body much darker than theirs, nearly approaching to a gloffy black. The part beyond the wings is divided into four joints, diftinguifhed into fo many rings; whereas the workers have but three, and thofe of a lighter colour. The more full of eggs, the more yellow is her belly. Her wings reach only to the third ring, but thofe of the workers extend to the end of their bodies. Her appearance is rather clumfy, but her deportment grave, ftately, and calm. She is armed with a fting fhorter than thofe of her fubjects. Its ufe is only to oppofe *rival queens*; for otherwife fhe will bear the rougheft handling, without attempt- ing to wound. She is very rarely to be feen, even with boxes of three windows; and, if by chance fhe is difcovered, inftantly retires from view.

Her

Her FECUNDITY is amazing; for in the course of a year she usually lays forty thousand eggs, or more: she has been seen to lay forty immediately one after another. Her body at the height of the laying season contains some thousands of eggs. If empty cells are not prepared, she is obliged to drop them. She is five times longer in laying a royal egg than a common one.

The *eggs* are little white bodies, fixed by their smaller end to the bottom of the cell. The royal cells are constructed on the edges or sides of the combs, (pl. II. fig. 4. k.) sometimes to the number of ten or twelve. These cells, when about half finished, resemble the cup of an inverted acorn, *c*, and are lengthened in proportion to the growth of the maggot or nymph. They hang in a perpendicular manner with the open end downwards, *c*. After the egg is deposited it remains in that state three days; and then being hatched, appears as a maggot in the shape of a half moon, lying at the bottom of the cell, surrounded with a clammy white substance, continually supplied by the workers for its nutriment. In five or six

B 2 days

days it grows confiderably larger, ceafes to take food, is then fealed up, *b*, with a waxen cap, and continues thus about twelve days, when the *royal nymph* burfts open the cover, and iffues forth a complete princefs. Cold weather makes two or three days difference in the time of exclufion. The queen is impregnated about Auguft, by virtue of which fhe is enabled to breed in the fpring, till fhe produces frefh drones.

Similar to the procefs above, is that of the *drones* and *workers*; excepting that the *eggs* are hatched in the common cells, which ferve in a double capacity, either for honey or brood. The cells for drones are generally the two middlemoft combs of the hive; the cells are deeper than thofe for the workers, and, when they happen not to be long enough, are lengthened by a cap of wax. They are generally hatched in twenty-one days.

DRONES

ARE thofe large bees (pl. II. fig. 5.) which ufually appear before the rifing of fwarms. They are the ONLY MALES, and

are

are larger than the workers; of a clumfy
fhape, and their extremity large, as are their
eyes; their trunk, or probofcis, fhort and
thin, and the body more hairy, They
make a much louder and rougher noife than
the workers; and having no fting, nor
inftrument to collect honey, are fuftained
by that of the hive.

It feems clear to me, that the drones are
of NO OTHER ufe but that of *propagation*.
I have, indeed, often found, that ftocks will
fwarm before any drones appeared; yet,
perhaps, fome were bred long before, re-
fiding in the warmeft part of the hive: and
which facts proved true; for drone nymphs
have been caft out in early fpring. Soon
after honey-gathering eeafes, they become
devoid of the fpermatic milky liquor, and
therefore are difcarded. The queen, containing fome thoufand eggs at a time in her
body, demands a larger fupply of the *prolific
juice* than a few drones are equal to furnifh.
This accounts for the large number of
drones found in the hives, as being abfolutely neceffary. As foon as the queen
finds *no* occafion for their fervice, they

feparate

feparate from the workers to the fides of the outward combs.

They are little noticed by the workers, and if killed at the doors of the hives do not refent it. Thofe that happen to remain in the ftocks till the cold weather arrives, foon perifh by it.

As their agency in generation, or, indeed, their utility at all, is ftill difputed, it is worth notice, they are endowed with a large quantity of a whitifh liquor in fummer, which the workers are fond of licking when a drone is fqueezed. The many thoufand times I have obferved drones in the combs, I never beheld one with its *tail* in a cell.

WORKERS.

THE *common bees, or workers,* (pl. II. fig. 7.) live about a year, but are very liable to premature death, by hard labour, high winds, birds, and by many other accidents. They are of neither fex, but abfolutely neuters. The young bees are diftinguifhable from the old, by being of a lighter brown. They are not all of one fize, a few being fhorter than the others, by being

hatched

hatched in shorter cells; but the dimensions of a cell cannot alter the sexual parts, only as to *size*, and not the male organs into female.

Their labour seems to be *indiscriminate:* they build the combs, nurse and sustain the young, collect honey, and defend the hive against all invaders. For cleanliness they are remarkable; have a quick and extensive smell, either for honey or honey-dew; but are not disgusted with many odours offensive to us, as paint, tar, urine, &c. partaking sometimes of such substances as are pernicious to them. Foreseeing impending storms, they make a precipitate retreat in great multitudes.

When *first* placed in a hive they work night and day, taking repose by turns, and sleeping in clusters. They can readily distinguish the bees of their own hive from all others; and highly resent the killing, or even disturbing, any bees of the same apiary, with vengeance attacking the aggressor.

As probably the following *novel* and curious discoveries may be pleasing to naturalists, their insertion will not offend practitioners.

CHAP.

CHAP. II.

DISCOVERIES.

UNEXPECTEDLY I *saw a queen* on a comb, near the window of a double box; the next day I was favoured with a like *view*; she remained each day about an hour; the bees very respectfully making a free passage for her as she approached. About a dozen of them tenderly licked and brushed her all over, while others attended to feed her.

During this interview I perceived several *eggs drop from her*, which the workers took no notice of. The box in which she then appeared was a *super* one; the under one had only *three* bars, and *four* apertures. The super-box seemed quite full of honey and brood. The queen tarrying and not choosing to descend, being obstructed by the middle bar, probably was the occasion of this reluctance; as also that of the bees from working in the empty nadir box.

From

From feveral fimilar difappointments I fur-
mifed, that the fcantinefs of the opening
for communication was the fole caufe. Inftead of *three* bars, from that time my boxes
were altered to *fix*, which *fucceeded* to my
utmoft wifh.

Another time I faw the workers very bufy
in demolifhing a ROYAL CELL, clofe to the
window of a box. It had been fealed up
fome days: but continuing fo beyond the
ufual period of exclufion, I fufpected fome
mifchance, and therefore was very intent to
obferve the refult. At five o'clock one
morning, the workers were very deeply
engaged in opening the fide of the cell: in
about two hours they had made a chafm
large enough to fee the nymph, and which
they were endeavouring to pull out, but in
vain. They then proceeded to a further
enlargement; when the *queen*, with hafty
fteps, and anxious looks, as if angry at the
delay, began herfelf the arduous tafk, the
workers remaining quiet fpectators. The
queen made feveral violent tugs to difengage
it, but her efforts proved fruitlefs. She then
retired,

retired, not without an appearance of difpleafure.

The workers then renewed their attempts, about a dozen at a time, and at intervals ceafed to enlarge, while they tried to pull the nymph out, but were ftill difappointed; for on pulling the nymph upwards fhe was preffed more into the convexity of the top. Four hours were thus employed; when the queen returned, with like demeanour as before, and proceeded with redoubled efforts to extricate the nymph; but ftill, unfortunately, with no better fuccefs, and finally relinquifhed the toil with great concern. However, the labourers refumed the tafk of enlargement from top to bottom, which was not effected till near twelve o'clock; a bufinefs of feven hours to draw the nymph out. It was full grown, but—*dead!* The feafon having been bad, the wax which compofed the cell was coarfe, and much thicker than ufual, fo as to render it impoffible that the young lady fhould extricate herfelf in due time.

During the time of the above obfervation, I beheld,

I beheld, in fome other boxes, royal nymphs burfting open the lower end of their cells, and inftantly iffuing without affiftance.

After many effays by various means, I never could procure a complete view of an intercourfe between a *queen and a drone*; but have feveral times been witnefs to thofe amorous preludes recorded by *Reaumur*. By confining a queen and a drone under a glafs tumbler, after fome little time the queen began to carefs the drone, frequently repeating fuch wanton geftures as would ftimulate a torpedo, or any other *male* but a *drone!*

Reaumur's relation of this myfterious affair ftates the refult of the royal embrace to be the *death* of the drone. The drones knowing, perhaps, this to be the confequence when *fingly* employed, may be the caufe of their extreme relu&ance. This, together with the violence ufed during their captivity, and the coldnefs of their fituation compared to the warmth of the hive, feems to account for the non-performance of *that* which naturalifts are fo defirous of difcovering.

I have feveral times placed two queens,

taken

taken from feparate hives, under a tumbler-
glafs, and immediately a *royal duel* enfued,
terminating in the death of both.

CHAP. III.

OF THE BEE'S STING.

THE ftinging of bees is often not only
painful, but has fometimes proved fatal to
man and beaft. Having frequently fuffered
under the *fmart*, it has taught me an ex-
perimental treatment of the wound.

Bees at a diftance from their hives, and
while purfuing their labours, are harmlefs
and peaceable; but if difturbed near their habitation, by hammering, buftling, or any
other great noife, or by ftanding before their
hives when very bufy, thefe intrufions will
urge them to refentment.

On thefe occafions the *face* is their chief
aim, particularly the eyes. In fuch cafes,
cover the face with the hands fpread, and
make a fpeedy retreat: they will not at that
time fting the hands.

During their active feafon, *gardeners* fhould do their requifite bufinefs near them early in the morning, or in the evening when the bees are retired to reft.

High winds very much difconcert and hinder their labours, and make them very irritable, and prone to affault any perfon that comes near their dwelling; and more fo, if it is at the time of their being anxious to fwarm, and if they are by fome means delayed therefrom.

To fome perfons they have a natural averfion, however unoffending, or however they may change their drefs, or though at twenty or thirty yards diftance.

A fingle bee will fometimes fly into a room, and fettle upon the hands, face, or neck; but they have no hoftile intent, and will prefently fly off again without wounding; provided no part of the apparel preffes upon them. They may gently be ftruck off, and they will fly out of the window.

The venom of their ftings is much ftronger in fummer than in winter. When a bee gets entangled in the hair, the alarm is great, but danger none, if the patient is
entirely

entirely paffive, till another perfon fearches for it, and, when found, crufhes it between his finger and thumb.

When bees have been a *little difturbed*, numbers will fly about a perfon near them, and with angry found (well known to apiators) warn them to depart, or they will fting. Retreat in hafte, covering the face with the hands, till the head can be protected among bufhes, or in fome dark apartment ; and there remain, till the violence of their fury is abated. It is very wrong, when a perfon is befet with bees, to ftrike, or buffet them ; for this is of no ufe, but will make them ten times more furious, and provoke multitudes to affift in the fray. Patience, and a fpeedy retreat, and fprinkling water over them that remain, are the beft expedients to get rid of them, which in about half an hour will be effected : but if any remain on the clothes, they may be brufhed off; except thofe on the face and hands, for that will make them immediately fting. Let them alone, they will quit of themfelves, when the reft are departed. If many continue to fly about, let water be thrown among them,

or

or blow them forth with a bellows, which they will fuffer without refentment. The fmoke of damp ftraw, or rags, will drive them away foon.

But the higheft degree of their rage is provoked by the *moving, fhaking*, or *tumbling down* of their hives; for then the whole army will rife in a mafs, and fall upon the aggreffor, be it man or beaft, hog or dog, to the imminent danger of the creature's life. Immerfion in water is the quickeft method to get rid of them, if any ponds, &c. are near. But if that cannot be conveniently done; taking refuge in a dark room, or out-houfe, and ufing the other means above directed, will be the moft likely to fucceed, till medical help can be procured.

REMEDIES.

NUMBERLESS have been the remedies propofed, and tried, without being generally beneficial. Thofe which have proved falutary to fome, were the reverfe to others; conftitutions and the fluids being infinitely various.

Some

Some are affected only in a fmall degree by a fingle fting; while others (though few) hardly at all, though by many. Again, many that are delicate and tender fuffer feverely, though ftung but flightly : thofe alfo who are of an *irritable* conftitution like that of the bees, fuffer to a high degree.

In a *curative* point of view, it is of the firft importance that a remedy be at hand, fo that it may be applied *immediately*, before the fubtilty of the venom gets into the circulation. After that happens, the medicine can but have a partial or weak effect. I have generally experienced my own *faliva* (fpittle) to be more beneficial than more pompous chemicals or galenicals (I fuppofe, chiefly, from its being always ready); rubbing it on the wound, tranfverfely from the direction of the veins, and not up and down; for that forces the venom more into the circulation.

A *fecond* remedy from which great benefit has been found, is, Extract of faturn, half an ounce; volatile alkaline fpirit, half an ounce; two drachms linfeed oil; fhake the extract and the fpirit well together, and then
the

the oil: it muſt be rubbed on the wound well, and conſtantly, as long as any pain is felt. It is dangerous if taken internally.

The *third* is dulcified ſpirit of ſal ammo‑ niac; adding one third of water, both being well ſhaken together. This has been found more generally efficacious than the preced‑ ing. It will not always prevent ſome degree of ſwelling, but ſoon aſſuages pain. It is of a harmleſs quality, and I have often uſed it about the eyes, without prejudice. To ſome, dulcified ſpirit of nitre has proved of preſent relief. Any of the articles may be had of the chemiſts, or apothecaries, at a cheap rate.

On great *emergencies*, if, unfortunately, none of thoſe medicines are at hand, com‑ mon linſeed oil ſhould be rubbed on the part ſtung: or in want of that, neat's foot oil, freſh butter, or hog's lard ſhould be ap‑ plied without delay, or the cure will be *re‑ tarded*, with an increaſe of danger, if the ſtings have been numerous.

In the mean time, tea made of balm, elder flowers, or lime tree flowers, or water gruel with a little ſalt-petre diſſolved therein,

C ſhould

fhould be prepared ; of which the patient fhould drink plentifully, and often ; refraining from all folid food, particularly that which is falted, or dried ; as alfo from acrid, acid, or fpicy articles. If a fever fhould intervene, James's powders give admirable relief. But if there is imminent danger, *medical* affiftance fhould be called in. Where the fymptoms are favourable, the tumours will gradually fubfide in a few days, without further applications.

The *like cooling* treatment is alfo to be ufed for *horfes, cattle,* &c. by enlarging the quantities, by mafhes, and by keeping them moderately warm in the ftable.

From the foregoing obfervations, perfons may juftly conclude, that thofe to whom the ftings of bees are very afflictive, fhould not, in common prudence, *attempt* the office of an apiator, nor approach bees, deftitute of a proper *drefs.*

Nor is it advifable to employ *fervants* about bees, that have a diflike to the bufinefs ; for, otherwife, it is a great chance but they neglect, or injudicioufly and perhaps fpitefully treat them.

CHAP.

CHAP. IV.

THE BEE DRESS

Is to be made of thin *boulting cloth*, which may be bought at about fixpence a-yard. It is to be fewed to the *brim* of an old hat, when reduced to two inches and a half in width; the cloth is to hang down a foot in breadth all round the head. A broad tape is to be prepared, long enough to tie the cloth, *clofe* round the neck, under the chin. But as the nofe, chin, and neck, would be liable to be *ftung* through the mefhes, therefore, to fecure thofe parts, fome *oiled linen* muft be ftitched oppofite the face and neck, within fide, leaving two inches and a half *free*, oppofite the eyes.

Or, a kind of hood of the like cloth may be made of fuch a breadth, that from the bottom of the *crown* of any *hat* in ufe, it may hang a foot below the rim. It is to be gathered up to a ferret binding, to let the crown through, and encircle it clofe round.

The

The portion which hangs down, is to tie round the neck, as before mentioned. Something for the mouth to grafp will be proper in both cafes, to keep the mefh at a requifite diftance. This laft hood is calculated to carry in the pocket.

The *oiled linen* is prepared by foaking linen in linfeed oil, and then fqueezing the fuperfluous oil out, and drying it in the air: this procefs will take two or three weeks. The procefs is then to be a fecond time repeated. Gloves made of it, though thin, will be impenetrable to the fting of the bees: indeed they will not attempt it. Garments made of it will effectually refift *wet*. The oil may be previoufly coloured by the ufual pigments, for green, blue, yellow, &c.

Befides the hood, a thick pair of tanned leather *gloves* will be neceffary, or other leather oiled only *once :* a portion of old ftockings is to be fewed to the extremities to draw tight over the cuffs of the coat. The *legs* muft be defended by a thick pair of yarn ftockings, drawn over thofe in common wear. The greateft care muft be ufed in putting on the hood, that no hollows or chafms be left under the chin, or about the neck; and for better fecurity, it will be proper to tie a handkerchief over the gathering round the neck, befide that of the tape. An apron before will be ufeful to prevent thefe prying infeds from tickling the belly.

Thus *apparelled*, *defiance* may be given to millions of bees, or wafps, and *all the operations* may be executed without dread or danger. Or if, by accident, hives are thrown down by cattle, hogs, &c. and the bees enraged; having this drefs on, the creatures may be affifted and the hives replaced.

Women fhould not meddle with bees, without this bee-drefs; nor then, without the addition of a man's coat, and I had almoft faid breeches alfo.

CHAP.

CHAP. V.

ON THE APIARY.

THE properest situation for an *apiary* is one exposed to the wind as little as possible ; it being detrimental, and proving often fatal to numbers of bees, by blowing them down, or into the water, or overturning the hives. Trees, high hedges, or fences, on the back and western side of the hives, will be necessary, to screen them from the violence of its force. But they should have a free opening in their front to the *south*, or rather south-east aspect. A valley is preferable to high grounds to favour their increase.

The hives should be well *secured* against hogs, or other creatures, which might displace the stocks, or otherwise disturb the bees, and injure themselves. Let the hives be set as near the dwelling-house as conveniently can be, or to rooms the most occupied, for the readier discovery of rising swarms,

or

or to be apprized of accidents. Befides, the bees habituated to the fight of the family, will become lefs ferocious, and more tractable; while the buildings will afford a protection from the wind and cold. The hives muft be clear of the dripping of trees, nor fhould long grafs, weeds, or dunghills be fuffered near them, as harbouring myriads of infects and vermin, that will prey upon the bees and their production. Neither are rivers, ponds, or large tubs of water eligible to be near an apiary, as great numbers will be blown therein.

It is very *wrong* to place *hives* on benches, which is always the fource of miftakes, quarrels, and often flaughter, by their interference with one another. A ftill worfe contrivance is that of little *cots*, or fheds, with fhelves therein, one above another; affording a greater harbour for their enemies, very inconvenient for the management, and indeed *impracticable* in the ftory method.

The *arrangement* I would recommend, is, that of *feparate ftands* for each hive, made by driving four ftrong ftakes into the ground, at equal diftances, as thus, : : correfponding

to

to the dimenfions of the hive *floors*, to reft
thereon : they are to be fixteen inches above
the earth, and the tops to be upon a level
with each other.

The ftands fhould be three or four feet
diftant from one another, and from any
wall or fence, in uniform *rows*, for the apiator's conveniency of managing each ftock ;
nor fhould the hives be fet higher than fixteen inches, in the ftory method ; for then
their height would be attended with many
difficulties.   Where perfons have many
ftocks, it is better to divide them into feveral
gardens, as being too numerous in one, frequently occafions quarrels ; eight or ten in
one place are enough.

Water is neceffary near an apiary in a
long *feafon of dry weather*. Put the water in
a broad difh, covered with fmall ftones, or
duck-weed, to affift the bees in drinking,
without wetting their wings, or being
drowned.

In *very windy* fituations, efpecially near
the *fea*, or great rivers, numerous bees are
deftroyed, by being blown therein, and
others very much injured and hindered, by
                                         being

being drove with violence to the ground, or other hard fubftances, with the lofs of their farina, fo laborioufly obtained,

Some have thought that an apiary near the *fea coaft* would be abundantly productive, by reafon of the bees being fond of fea-water. This point I have made obfervation on, my refidence being only four miles from the fhore, but could not perceive that the bees fhowed any fuch partiality, unlefs neceffitated by a long feafon of very hot and dry weather. Nor did they much affect the wild thyme that grew on the fand-hills adjoining; nor are they fond of falt.

CHAP.

CHAP. VI.

ON PURCHASING OF BEES.

THE beſt time for eſtabliſhing an apiary, is juſt before the taking up ſeaſon : which is *generally* about the latter end of Auguſt, for then bee-keepers reſerve as many of the beſt ſtocks as they judge expedient for their next ſummer's ſupply ; and, therefore, after that period are not diſpoſed to part with any, unleſs at an advanced price : whereas, by purchaſing ſome time before, a choice may be made of the *beſt*, and at the accuſtomed rate.

They ſhould be ſelected by a *ſkilful* perſon, in a cool evening, or rather morning very early. By tapping about the hive, a pretty near gueſs may be formed, whether or not it is full of bees, as alſo if full of combs. But for greater certainty, turn thoſe that *ſeem* heavy upon the edge of the hive, and obſerve if the *interſtices* between

the

the combs are crowded with bees, and the combs worked down to the floor. If white, or of a light yellow, it denotes their being of the present year's produce, and fit for the purpose; but if they are of a very deep yellow, or brown, they are of the last season, and not so proper; while those that are dingy, or *blackish*, are *old*, and wholly unfit to furnish a prosperous apiary. To avoid deception, observe, that though a hive may have the *edges* of the combs of a light yellow, they may be old stocks neverthelefs, whose combs the preceding year not having been completed, have in the present had *new* borders added to them of virgin wax, so as to look like young stocks. Look carefully *between* the combs, as far as the bees will admit; and if the interior parts appear favourable, form a judgment accordingly. The hive should be poifed in the hand; and if it be about half-bushel fize, and weigh twenty-five pounds or upwards, it is another teft of its being a good stock. But the weight alone, of old stocks, cannot be relied on, as great part of the combs may be crammed

with

with old farina, and other impurities, as
mentioned hereafter.

One good ſtock bought at the proper
time, is worth *two ſwarms* bought in the
ſpring; for ſuch a ſtock will ſwarm once or
twice, or yield *two* or *three* hives full of
honey; whereas, from a ſwarm, little or no
profit can be expected the *firſt* year.

But ſhould the proper ſeaſon have been
neglected, a *prime or firſt ſwarm* ſhould be
ſought, at leaſt large enough, in common
ſituations, to fill a peck, and if a good one,
half a buſhel. Small ſwarms will turn to
little account, and balk the expectation.

The ſwarm is to be brought home in the
evening of the day it riſes. If a large one
cannot be had among the neighbours, *two
or three* may be united, to form a powerful ſtock.

If a ſwarm is delayed being brought
home for two or three days, portions of
combs will have been conſtructed, which
may probably be diſplaced in the removal,
with the bees thereon, and may be damaged, or cruſhed, and ſo be the ruin of
the

fwarm : to avoid which, let it be removed
at day-break.

To *transfer* the fwarm from the common
hive, into one of your own, or into a box,
invert that which has the fwarm in a pail,
bucket, or the like; lay two thin flat flicks
acrofs, and then fet the empty hive over it;
ftop the juncture with a cloth, and before
morning the bees will have afcended into
the upper one. But if not, let them ftand a
day longer; when, if they ftill are reluctant,
ftop the juncture quite, and beat round the
lower hive with two fmall flicks, till they
afcend, which may be known by the great
buz in the upper hive.

Or, as foon as two fwarms are brought
home, fpread a cloth on the ground, and
lay a flick acrofs : then ftrike the edge of the
hive with violence on the ground; the bees
will fall out in a lump : then take the other
fwarm, and ferve them in the fame manner,
clofe by the firft; fet an empty hive over
them, refting one edge on the flick, and
cover them with a cloth. If they are found
to quarrel when afcended, they muft be
fumed as directed hereafter.

REMOVING

REMOVING OF STOCKS ſhould be in the evening, or very early in the morning. The hive ſhould be raiſed by three or four wedges, ſome hours before, provided the floor is *not moveable ;* or *otherwiſe* many bees will remain on the floor at the time, and be very troubleſome.

A cloth muſt be laid on the ground behind the hive to be removed; nimbly lift the hive thereon, and, gathering the four corners tight, tie them faſt on the top : immediately draw a ſtring cloſe round the body of the hive, to prevent any bees crawling between.

If they are to be carried a conſiderable diſtance, they may be reſted on the ground, as occaſion may require. Hand barrows, or yokes, with a hive ſuſpended at each end, or a long pole on men's ſhoulders, and a hive or two between, may be advantageouſly uſed for their conveyance.

But when it is for ſeveral miles, a coach, or cart with plenty of ſtraw at the bottom, to break the ſhocks of the carriage, and then proceeding with the ſloweſt pace, and taking the cool of the morning, will prove

I   a ſafe

a safe and convenient removal. If any of the combs should, however, be broken, and fallen on the cloth, when the hive is taken off, let them remain thereon, and set the hive in the place or stand designed for it; and gently spreading the cloth with the bees on it on the top, by the morning they will have quitted, and entered by the door of the hive.

A stock should not be set *close* to the bee-house front, the first night of its being brought home, that the straggling bees may find their way into the hive by the door, and then no bees will be crushed. Straw-hives, being of a circular form, leave a considerable vacancy between the hive doors and front, which *next* night must be stopped, by thrusting part of a hay band, or clay, or stiff cow-dung, to fill the chasms, but leaving the door-way free.

Purchased swarms in spring, on bringing home, are to be immediately set on empty hives; and thus, by being doubled at *first*, will save that trouble afterwards.

, CHAP.

CHAP. VII.

ON THE FORMATION OF STRAW HIVES.

STRAW is the beſt *material* for hives, as
beſt protecting the bees in the extremes of
cold and heat, and alſo generally eaſieſt
to be procured. Where it is not ſo, ruſhes,
wicker-work plaſtered over, or ſedges, muſt
be ſubſtituted.

Of ſtraw, *unthraſhed* RYE is preferable, as
thraſhing ſhivers the ſtraw, and makes it
rough and ſhaggy, which the bees with
much labour are obliged to gnaw off. My
hive-maker laid the ſtraw in a chaff box,
and ſo readily cut off the ears.

The PLAN I propoſe is, THREE HIVES
to each ſtock. The ſize I have found moſt
convenient is that of half a buſhel: larger
are very *inconvenient* to manage; while theſe,
by *ſtorifying*, give ample room for all that
the bees can want, at the ſame time admitting triplets to be taken off the ſooner.

They are to be *nine* inches high, and
twelve

Plate 1

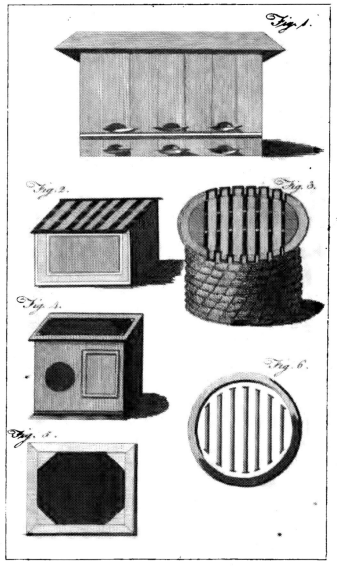

Fig. 1.

Fig. 2.

Fig. 3.

Fig. 4.

Fig. 6.

Fig. 5.

Plate II.

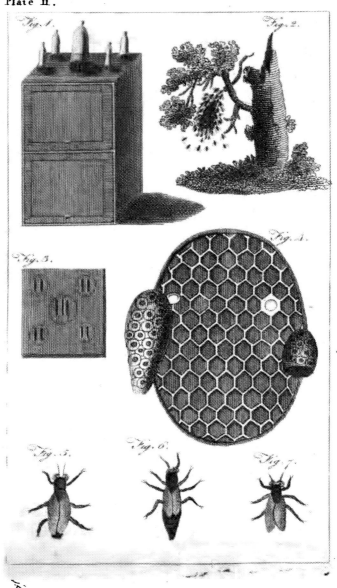

twelve wide, in the clear, on the infide, i. e. exclufive of the *top*, (pl. 2, fig. 3.) . The *body* is to have no ftraw top *fixed*, or worked to it, as in common, but is to be a feparate piece. The body of the hive, therefore, re-fembles a broad hoop; and, like that, muft be perpendicular, or ftraight down; and not one part *fwelling*, or being wider than ano-ther.

The ftraw COVER is to be made *quite flat*, like a round mat, but wide enough to extend an inch beyond the edge of the hive. There needs only one cover to three hives. The greateft proof of the maker's fkill will con-fift in his exactly following the prefcribed dimenfions, and in the evennefs of his work; particularly in both edges, that they may admit one hive being fet on another, with-out any chafms, and that *promifcuoufly*, or hab nab.

In one of the edges a diftance of full three inches is to be left *free* of binding, for a *door-way*. But a more proper one may be formed by a fmall piece of wood, four or five inches long, in which a door-way is to be cut, of three inches long, and *three-eighths*

D of

of an inch in height, and worked into the round of ftraw.

Or, what will be ftill better, is to take a rod of willow, or hazel, while green, and bend it to a circle of a proper fize for the hive. When it is wanted, reduce it fo as to have two flat and even fides; cut a proper door-way out, and burn holes at due diftances to receive the brier binding, by which the firft round of ftraw is to be faftened to it. If the binding is carried wholly round the hoop, the binding will be foon rotted by the wet, and prove of little more fervice than if there had been none; but otherwife it will preferve the hive much longer, and be more convenient in many refpects.

As foon as hives are made, they fhould be fet feparate on *level* boards, or the like, and another on the top, and heavy ftones laid on them; but firft a perfon fhould jump upon the boards to reduce the edges to a proper evennefs. This practice muft not be neglected.

Befides the flat ftraw cover, *all* the hives muft have WOODEN TOPS, (pl. 1, fig. 6.) to make which, procure a board of the width

2   of

of the hive, and half an inch thick, free from knobs. *Seven* fpaces or openings are to be cut, b, b, b, b, b, b, b; each exactly half an inch wide; the *length* of the three innermoft, *eleven* inches; the two next, *nine*; and the two outermoft, *fix* inches. The carpenter muft be attentive not to deviate from thefe directions in the fmalleft degree, as a trifling neglect will render the whole ufelefs.

In cafe boards of a proper width are not to be had, one ten inches wide may be fubftituted, braiding circular pieces on the fides after the top is cut out, to fill up the deficiency. Round the edges a hoop of tin, or flight ozier, muft be tacked to ftrengthen it, and prevent its fplitting. A long braid or peg fhould pafs through the fore and hind parts, and enter the edge of the hive, to keep the top from being difplaced; taking care that the heads of the braids are driven rather *below* the furface of the wood.

A CHEAPER TOP may be made of narrow flips of wood, which I name BARS, *fix* in number (pl. 1, fig. 3. a, a, a, a, a, a); defigned to be laid acrofs the top of the hive, at half an inch diftance from each other; the

two

two outermoft bars to be one inch and a
quarter wide, and the others one inch and a
half. Two flips of wood, b, b, an inch wide,
are to be braided acrofs the bars within fide
(or rather let in, to be flufh on both fides)
near the ends, to faften them together, and to
keep them at their due diftance. The crofs
pieces will thus be below the edge of the hive,
while the *ends* reft on it: But fince the breadth
of this *frame of bars* will not be quite that of
the hive, the deficiency muft be fupplied by
two fmall circular pieces braided on the edge
of the hive, leaving two half-inch openings
between them and the bars. As the ends
of the bars, when laid on the hive, will
leave vacancies between, thefe muft be ftopped by cow-dung of a due temper, which,
when dry, will be fufficiently tenacious.
Take care that the whole top be even
and fmooth. It fhould be laid on always
in the direction of *front* and *back*.

The ftraw covers are to be faftened on by
loops of cord, or rather leathern thongs, paffed within, at about two inches below the
top of the hive. They are to be four in
number, placed at equal diftances, and a cord

to each pair, to draw them tight over the
top,

The HIVE FLOORS fhould be one inch
thick, of yellow deal planed on one fide
only, truly level, and of fixteen inches diameter. Where boards of that width are not
eafily to be procured, an additional piece
muft be *rabbeted* and *dowveled* to it. Two
crofs pieces are to be nailed underneath, to
ftrengthen and prevent its warping; or rather they fhould be nailed upon the ends.
Three of the corners may be cut off, leaving
the *fourth* for a place to alight on. ONE
floor only is requifite to every *three* hives;
but two or three *fpare ones* will be convenient on many occafions.

COTTAGERS may contrive *tops* from thofe
cuttings of trees which are ftraight, of an
equal thicknefs, and of a length as above
defcribed. Thefe, while green, may be eafily cut flat, with a knife, of the proper meafure, by firft laying them over the top of
the hive, at the diftance of half an inch from
each other; they may then be marked, and
cut to their juft length. Two pieces are to
be braided under their ends, fo as *not* to pre-

vent

vent the crofs pieces from finking into the infide; and to hold the bars fteady, without fliding backward or forward. The vacancies between the bars on the edge of the hive are to be filled up with cow-dung, which, when dry, will be fufficiently tenacious. Care fhould be taken to make every part of the top fmooth and level; which if not fo, reduce it by laying heavy weights thereon.

Hive-makers in fome places have affected confiderable difficulty in making hives of the *form* I have prefcribed, but without juft grounds: the perfon employed by me, after a little practice, could make them as expeditioufly and eafy as thofe of the common fort.

His *method* was to make a common hive, the circumference of whofe bottom was exactly to the dimenfions I defired; on the edge of this he worked a round and a half of ftraw, bound on with a *cord*, and then continued to proceed with brier binding, having by him a ftraight ftick, of the due width, as a gauge, and to keep the work truly perpendicular, or upright. If the hoop I before mentioned is provided for the bottom

tom edges of hives, the work might be begun
and carried on from *that*.

When he had got about half the intended
width, he finifhed the *round even*. Then
loofing the cord from the part he began at,
that part was taken off and inverted, and the
round left loofe by the cord was re-bound
by brier: and thus he proceeded till he completed it. It is to be noticed, that the *part*
firft begun at was in the middle when
finifhed.

Apiators who underftand what I have
written on this head, fhould offer a
good price to thofe who are reluctant in
making thefe hives, and fhould ftand by
while the workman endeavours to make one;
and by giving occafional directions it may
be eafily effected, and they may be introduced over the kingdom.

It will be a good method to plafter one
fide of the ftraw top with cow-dung, even
and level, which will prove more eligible in
introducing the fliders.

HACKELS or COPPETS are made of wheat-
en ftraw. The method is this: Take a fheaf,
bind it with a cord ten or twelve inches

D 4   below

below the ears : with the left hand gripe a
fmall parcel or locket (about 60 ftraws) of
the part above the cord, and with the other
hand a like locket; and giving it a twift round
the firft locket, bring it down clofe to the
cord, pulling the other locket ftraight down.
Take a third locket and twift over the preceding ; and thus continue to twift and turn
down until the whole is finifhed, except
three locks, one of which is to be brought
between the other two, which are to be tied
in a knot over it. Then reducing the whole
as flat as can be, run a fhort forked ftick
through the knot, to prevent its ftarting.
The hackel may be made in about twenty
minutes.

This form is the beft fuited to the purpofe
of any that I have.feen ; they fit clofe to the
top of the hives, keeping them warmer and
drier, which is of great advantage in winter
and fpring. Neither are they fo liable to be
blown off. The part before the doors fhould
be clipped fo as to admit the fun's rays. For
fear of ftorms, a hoop may be thrown over
them, and faftened by two ftrong fticks with
crooks at their ends, and thruft into the
ground

ground on each fide. This will be a good fecurity at all times.

Placing the hives at the diftance before ftated, will preferve the bees from quarrelling, or emigrating from one hive to another.

Opulent perfons, to whom the appearance of ftraw hives may feem inelegant, might have them concealed from view by fuch fhrubs as are of fervice to the bees, planted at fuch a diftance as not to intercept the funfhine to the front of the hives.

Or, handfome covers, fomething in the fhape of hackels, terminating in a point at top, and painted, would have a pleafing appearance.

Or, a SCREEN in perfpective, of rocks or ruins, &c. with proper openings for the bees to iffue from behind, on floors properly difpofed, on which they fhould be placed as in a bee-houfe.

N. B. By *ftraw covers* are not meant TOPS, which are of wood, with bars. Nor are *hackels* meant by the term *tops*.

CHAP.

CHAP. VIII.

ON BEE BOXES.

BEE boxes are beſt made of ſeaſoned
yellow deal, free from knots, and one inch
thick.　The boxes are to be *ten* inches high,
and *twelve* ſquare; clear in the inſide (pl.
1, fig. 2.)　One of the ſides is to have a
pane of *glaſs*, *d*, of the whole width, and ſix
inches in height, with a ſhutter half an inch
thick, to be let into a bevel at top, and reſt
on a ledge at bottom, and to faſten with a
button, *a*; this is to be eſteemed the *back*.
There muſt be a *door-way* in the bottom edge
of the front, four inches long, and five-
eighths in height, excluſive of the *threſhold*,
which is to be one-eighth of an inch thick,
to be *let* into the edge of the box, and on a
level therewith.

A ſlip of wood is to be fitted for a door,
to turn outward to the left, on a pivot or
pin, and to ſhut in a bevel, with a ſmall
notch, that it may be opened by the point of
a fork,

a fork. It muſt ſhut ſo far in as to be fluſh with the ſide of the box.

The TOP (pl. 1, fig. 2.) is to be compoſed of *ſix* ſlips of wood, which I name BARS, a, a, a, a, a, a, three quarters of an inch thick; the *two* outermoſt, one inch and a quarter broad; the other *four*, one and a half. The ends of the *ſecond* and *fifth* bars are to be let into the front and back edges of the box, and fluſh with the outſide; the remaining four bars are to be of a due length, to paſs eaſily *withinſide* from front to back. Two fillets, each an inch broad, are to be braided to the bars, or rather *let in* tranſverſely, of the diameter of the box, and near their ends, not only to keep the bars at half an inch exact diſtance from each other, and from the ſides of the box, but to connect the whole like a frame together, and to take *in* or *out*, with the combs fixed to them, at pleaſure. The bars (1ſt, 3d, 4th, and 6th) ſerve alſo to prevent the frame from ſlipping from its ſituation. The top, thus made, will have *ſix* bars, and *ſeven* apertures, or openings, like the ſtraw hives.

There is to be but ONE CLOSE COVER, or
lid

lid of wood, three quarters of an inch thick, to three boxes ; which is to take off and on by means of four fcrews, one at each cor-ner.

Loose floors are to be provided with the boxes, to be planed on one fide, and filleted at the ends to prevent warping, and of an inch more in their dimenfions than the tops of the boxes. If a board broad enough cannot be had, a leffer muft be added, *rab-beted* and *doweled* thereto. One floor *only* is neceffary for a fuit (three) of boxes, but two or three fpare ones will often be wanted.

OBSERVATIONS.

A minute exactnefs is abfolutely necef-fary in working the boxes; for though the unexperienced may imagine the deviation of a quarter or eighth of an inch from what has been directed will be of no confequence, neverthelefs fuch miftake or negligence in any part would render the apparatus *unfit* for the ufe it was intended for.

Firft obferve, that the edges of the boxes, both top and bottom, are to be truly level, that when indifcriminately fet one *over* or

under

under another, no chafms or vacancies are left between them.

Secondly, that the frame of bars be made to take out with eafe.

Thirdly, that the fcrews for the covers fhould be flight but long, to pafs in at the fides, exactly one inch and a half, from front and back, fo that any cover may fcrew on any box, without making frefh holes. They fhould always be greafed before they are put in, or they will become rufty, and not to be drawn out without great difturbance to the bees, and much inconveniency.

Fourthly, great care muft be taken that no fnags or fplints of wood, heads, or points of nails, rife in the leaft degree above the furface, as a brafs plate is defigned to flide over the tops.

A neceffary appendage, as well to the hives as boxes, are TWO BRASS PLATES, of one *fixteenth* of an inch thick as near as poffible, fifteen inches wide, and fifteen and a half long, which half inch is to be turned upright to pull it out by. They muft be fet on a true level. If they are *thicker*, the bees will efcape on their introduction; and if

thinner,

thinner, they will not be ftrong enough to retain their neceffary elafticity and level, but will bulge in the middle, and let the bees out.

The braziers or ironmongers will fupply them. I gave in London fixteen pence per pound, and they came to eight fhillings.

But as in many counties large brafs pans or kettles are ufed, and, when unfit for boiling ufe, are fold as old brafs; the bottoms of fuch of thefe as are of the proper dimenfions, and not having holes of a fize for a bee to pafs, will do better than new, as being tougher; and any fmith will reduce them to a level, and turn up one edge. They may be bought at the price of old brafs, i. e. about fix-pence per pound. A *pair* come only to four fhillings.

I had an iron plate made which came to near as much as the brafs, but did not keep its level fo well, and was more unhandy. Steel, being elaftic, would retain the level much better, but I fuppofe would be dearer, and liable to ruft; which brafs is not, and will at all times fetch a large fhare of its firft coft.

Eight

Eight or nine fhillings by fome may be thought too expenfive ; but the great *utility* and *conveniency* of the plates, I am warranted to fay, will much over-balance that increafe of price. Every apiator muft be *feelingly* convinced of the difficulty and embarraff-ment of feparating hives of bees, and in the other operations, by any of the methods made public. Indeed, by them the bees of *under* hives are prevented from affaulting the ope-rator ; yet thofe of the UPPER ones are left entirely FREE to execute their whole re-venge.

By the ufe of the two plates, or DIVIDERS, and by doors to fhut, this great danger and inconvenience is entirely avoided, as the bees of BOTH hives are EQUALLY in-clofed, and prevented from infulting the apiator.

Befides, if ONLY one is bought, it is adequate in advantage with any other con-trivance, and will fuit hives as well as boxes.

Moreover, the plates are not perifhable articles, but with care may laft for genera-tions; and it muft be remembered that the

charge

charge lies. on the whole apiary, and *that*
only for *once.*

COTTAGERS, whom I wish to benefit, or
others, may club in the purchase, by which
the cost will be but slightly felt. Or perhaps
country shopkeepers would find it their interest to be furnished with suits of plates to
let out.

I have proposed a large window to a box,
as I found a small one of little use, and affording but little entertainment. Those who
would choose a more enlarged view of the
bees in boxes, may have large windows in
the *three sides.*

DOORS to the *hives and boxes* will be found
of great advantage on many occasions, particularly in passing the dividers under hives,
to prevent the egress of the bees if the door-
ways are stopped, and on various other occasions.

Boxes of bees placed in the window of
a room much incommode the company
whenever the window is opened. The *side*
of the room suits better: a proper opening to be made in the wall, and a small tin
trough adapted to pass through to the door-
way

way of the box. On the inside a shelf is to
be fixed, that the box may stand so close
as to leave no admission for the bees into
the room, and be so secured as not to be
displaced by any carelessness or inadver-
tency.

Apiators who have boxes, but whose openings are on a different plan to that now of-
fered, may, at a little expence, have them
altered thereto, provided the dimensions of
the boxes do not exceed that of the dividers.
The superfluous vacancy may, however, be
filled up with solid wood, or new tops may
be made with the bars and apertures, as I
have described, though the box itself be
much larger. For should the dividers be
enlarged, the hands will not extend suffi-
ciently underneath to keep them close, or
steadily to support the great weight: therefore the apertures and bars must not be
longer than those of my plan, commencing
from the *back*. Octagon boxes may have a
section of the back taken off, and a large
window supply its place.

The timber of the boxes is directed to be

E one

one inch thick, for one quarter of an inch thinner will render them not warm enough.

The *floors* of the hives and boxes being *moveable*, will be of very great utility and advantage in all the operations, and muſt be ſo evident to every reflecting apiator, as to need no further recommendation.

CHAP. IX.

OF A BEE HOUSE FOR THREE STOCKS.

IT is to be formed as in pl. 1, fig. 1, and ſix feet long, excluſive of the poſts.

Four poſts of three inches ſquare.

Two long rails to nail the floor upon, and two ſlight ones to nail the roof to.

A floor, ſeventeen inches in breadth, to be laid acroſs the rails.

A roof, four boards.

Two folding doors.

The poſts are to be fixed to the due length, and ſeventeen inches in breadth to their out-

ſide

fide. They are to be fecured in the ground at a proper depth, and five feet above the earth, and fet truly perpendicular. The tops to be bevelled one inch and a half.

The two *ftrong rails* of an inch thicknefs are to let into the pofts on the outfide, and ftrongly nailed, one in front, the other behind : to thefe the floor is to be faftened, *crofs-wife*, perfectly level.

The *flight* rails are to be let into the tops of the pofts clofe to the ends of the bevel, to nail the roof upon.

On the bevel of the pofts are to be fixed *two boards*, each fix inches wide, to extend beyond them two inches behind and before.

Two more boards, each at leaft twelve inches wide, and one thick, of yellow deal, and free from knots, are to be nailed floping againft each other, to complete the roof. Their edges on both fides are to be bevelled off fo as to meet at top, and make a neat joint ; and to prevent warping, braces acrofs on the *infide* will be neceffary.

The FRONT of the houfe, *A*, is to be inclofed by three quarters of an inch boards,

E 2 placed

placed perpendicularly in lengths, from the top to the rail of the floor, and rabbeted to each other. The boxes are to ſtand ſix inches from the ends, and eight from each other.

Openings are to be cut againſt each door-way of the boxes, *ſix* inches in length, and *two* in depth, *a, a, a,* eſtimating from the loofe floors of the boxes.

Similar openings are to be cut eleven inches higher up, in a line with the firſt, and even with the tops of the boxes when their covers or lids are *off.*

To the edges of the openings circular pieces of wood are to be braided, a little declining, for the bees to alight upon.

A batten, bevelled at both edges, ſhould be nailed on the outſide, juſt under the higheſt alighting boards, to ſtrengthen the front boards, and prevent their warping or caſting.

The BACK is to have two doors, ſhutting againſt each other in a rabbet, and to faſten with a haſp.

The ends are to be cloſed as the proprietor chooſes.

Good

Good painting will be of advantage to
preferve the whole. The door-ways fhould
be of different colours, for the bees the
better to diftinguifh their refpective habitations.

OBSERVATIONS.

The junction of the boards at the top,
however clofe at firft, will gape afterwards
by the changes from heat to wet; to remedy
which ftop it with putty, or rather, as foon
as it is nailed on, a flip of thin lead, of two
inches broad, fhould be tacked over the junction, which will effectually prevent wet
from getting through. Bohea tea-cheft lead,
that which is *whole*, will anfwer the purpofe.
It is of the moft material confequence to
bees to exclude wet. I have tried feveral
other materials for roofs, but none anfwered
fo well; and mine is a very trying fituation.

If the front is not truly perpendicular, and
the floor truly level, the boxes will not fit
clofe to the front, and thereby leave vacancies

between

between by which the bees may pass into the house; which would be very detrimental.

The openings for the paſſage of the bees are *larger* than thoſe of the boxes, as being more convenient on many occaſions. No openings are made in the house for *triplets*, as being unneceſſary.

Three of the front boards of the house, in which the openings are to be cut, ſhould be eleven or twelve inches wide; or they will be too much weakened, by cutting ſix inches in length out, to ſtand true.

The *principal intention* of a bee house and boxes, is for the more commodious inſpection of the bees by the curious and wealthy. *Three* ſtocks anſwer this deſign as well as a larger number, as they furniſh only a repetition of the ſame ſcenes. However, a bee house is, in ſome reſpect, of *real use* to thoſe who keep a number of ſtraw-hived ſtocks, as STANDARDS, from which, by inſpection, a judgment may be formed of the good or bad condition of the ſtocks in ſtraw hives : but, that boxes are *more productive* than thoſe,

is

is a *great mistake*, if both are managed by the same method of STORIFYING.

Many contrivances for the purpose of sheltering boxes have been practised as a substitute for a house; but, in the end, are not cheaper, and not near so convenient for performing the operations; neither are they so eligible for inspection. My bee house here, ten feet long, cost me near thirty shillings,

E 4   CHAP.

CHAP. X.

ON STORIFYING.

OF all the methods which have hitherto
come to my knowledge for the conducting
of bees, that of *ftorifying* undoubtedly yields
much the greateft *profit*, and is the moft con-
genial to their natural habitude, and ftyle of
working.

By ftorifying is meant the fetting of one,
two, or three hives over each other, as *duplets*
or *triplets*.

It is found that *three* pecks of bees in one
hive, will collect more honey than a bufhel,
divided into *two*; becaufe a *fingle* hive has
not combs enough to receive the numerous
eggs that a queen is capable of furnifhing,
and cells fufficient at the fame time to
hold the honey.

Thus being limited to a fmall compafs,
the *increafe* muft proportionally be fo too.
For great part of the bees are neceffarily
employed in *rearing* the young, and therefore

fore the number of thofe who are occupied
in collecting honey is not near fo great as
has been imagined.

A good *florifier* that has not fwarmed, or
has had the fwarm returned, will increafe
thirty pounds in feven days, in a favourable
fituation and feafon : whereas a fingle-hived
ftock in the fame apiary and feafon, that
has fwarmed, will not increafe above five
pounds in the fame time. For every fwarm,
the leaft as well as the greateft, is provided
with a queen, equal in fecundity to the
queen of the largeft ftock ; and as the brood
fhe brings continually demands the labour
and attendance of probably near *half* the
bees ; this circumftance renders the other
moiety, from the *fmallnefs* of their *number*,
unable to accumulate a *large* quantity of
honey in the fhort time it moftly abounds.
Whereas, by doubling, and trebling the
hives, the bees are never at a ftand for room
to extend their combs, as faft as requifite for
honey or brood.

Bees, confidered individually, *live* about
a year, progreffively coming into birth, and
as gradually decaying. It hence follows,
that

that thofe born in autumn, or fpring, or in the intervening months, inevitably die about the fame time in the fucceeding periods of time, and fo in a regular proportion during the breeding feafon ; but this is not perceived while the brood is rapidly increafing, and counterbalancing the chafms made by death.

The queen often lays two or three hundred eggs in a few hours; which occafions as fudden a difappearance at the ftated period, and which accounts for that great *thinnefs* obfervable in hives after the fwarming feafon is over, as if a fwarm had efcaped, This likewife demonftrates, that at the *general* time of *deprivation*, all hives, or ftocks, according to their populoufnefs, are compofed of bees of all *ages*, from thofe in embryo, to thofe of old age. Confequently, although individuals die daily, young ones rife to birth, to fucceed them, as do the human race in towns and cities. But, by ftorifying, the family is perpetuated to any length of time; without the *cruel neceffity and trouble of deftroying indifcriminately both old and young.*

The

The ftory method can in no cafe be prejudicial, though the bees fhould be *prevented
thereby from fwarming:* on the contrary,
it would be a great advantage if it did fo;
for then artificial fwarming would not be
wanted to perpetuate ftocks, which would
be effected without fuch affiftance. Writers
have however followed each other, by af-
ferting that by ftorifying no fwarms will
rife. From long experience I am certain
of the reverfe. When duplets or triplets
do *not* fwarm, it is *not* from *that caufe:* it is
from abortions of the royal brood, and feveral other cafualties.

Nor is there any danger of being *over-
ftocked;* for however numerous a ftock may
be in bees during fummer, in winter they
will be reduced to a *quart.* Befides which,
bad feafons often happen, and many accidents arife that will require recruiting, and
which may be happily effected by forbear-
.ing to double a good ftock, and a fwarm will
be the fooner obtained. The following Estimate will fhow how far the advantage in-
clines to *ftorification.*

A Com-

A Comparative Estimate of Stocks kept in Single Hives, and those placed according to the Storifying Method.

FIRST YEAR. Dr.			FIRST YEAR. Cr.		
12 stocks on an ave- rage, yielding 15 lbs. of honey each, is 180 lbs. at 6d. £.4	10	0	12 stocks on an ave- rage will yield two additional hives of honey, of 16 lbs. each—384 lbs. at 6d. - £.9	12	0
Supposing each hive to have a cast, each of which usu- ally affords 3lbs.— 36 lbs. at 6d. 0	18	0	Wax, 1½ each hive, 1	16	0
Wax 1 lb. each, and 4 oz. the cast, at 18d. - 1	2	6	£.11	8	0
£.6	10	6	Discount for the *extraordinary* ex- pences, viz.		
N. B. They are sup- posed to emit 12 good swarms, to stand for stocks.			24 hives at 14d. 1	8	0
			12 floors, - 0	6	0
To balance in fa- vour of the story method 2	1	6	24 wooden tops, 0	12	0
			2 brass plates, 0	10	0
			£.2	16	0
£.8	12	0	£.8	12	0
Thus at the end of the year the stocks will be equal.					
SECOND YEAR.			SECOND YEAR.		
12 stocks being the last year's swarms £.4	10	0	12 stocks produce as last year £.9	12	0
Casts, or small swarms 0	18	0	Wax - 1	16	0
Wax, - 1	2	6			
£.6	10	6			
Balance in favour of storifying 4	17	6			
£.11	8	0	£.11	8	0

From hence it appears, that by laying out *two pounds fixteen fhillings* for the extraordinary *apparatus* of the *firft* year, a fuperior profit is to be gained of *two pounds* one fhilling and fixpence. But in the *fucceeding* years it will amount to *four pounds feventeen* fhillings, that is about fifty per cent. per annum, on the two pounds fixteen fhillings fo laid out: or four pounds feventeen fhillings and fixpence a-year more, gained by ftorifying *twelve* ftocks, than by a like number in *fingle* hives.

This ftatement is made upon the *loweft* calculation in favour of ftorifying, which ufually yields much more honey and wax than here affigned, and that greatly fuperior in quality, and confequently more valuable; but which *cannot* be obtained from *common fingle hives.* The inftruments are rated higher than what they will ufually coft, befides their advantage of durability.

Though I fuppofed each common-hived ftock to emit a good *firft* fwarm, which they often do not, or it is frequently loft, and though fome often afford two or three, they in general are but trifling, and abate confiderably siderably of the produce of the mother stock, often to its ruin—what I have allowed for casts, in the common run, will be the full amount.

The estimate is founded on the productions of *middling* situations; but in *better*, a single hive may produce a stock of from thirty to forty-six pounds weight, grofs; the *higher* likewise will be the proportional advantage in storifying. *Where* hives weigh so, they are usually much larger than the general size: and I think in the *single* method, no hive should be lefs than three pecks, or perhaps a bushel, but not more than twelve inches in height. The twelve stocks will require three shillings and sixpence to be laid out in new hives, every *third* year, which I set against twelve new hives at least, which must be bought for swarms in the single management. *No other branch of husbandry* (I am inclined to think) *will return so large an interest on so small an expenditure.*

Besides the advantages already mentioned, there are others of consequence which deserve notice. 1st. In avoiding the unnecef-

fary

fary and difagreeable trouble of SUFFOCAT-
ING the bees. 2d. In relieving fwarms
when too large. 3d. In preventing idlenefs
in their lying out. 4th. In uniting of
fwarms. 5th. In the means of cleanlinefs
and wholefomenefs. 6th. In preferving
them from moths, mice, and other infects,
by the frequent fhifting of the hives. 7th.
In giving ample and timely enlargement.
8th. In being provided againft bad feafons.
Laftly, In taking but little room in an apiary:
as for inftance, four ftocks will require no
more ground to ftand on than they had at
firft; while common hives will demand
twice or thrice as much for fwarms, but
producing lefs honey.

The INDICATIONS FOR STORIFYING
ftocks, are the appearance of an increafe of
numbers, and in their activity, favoured by
the mildnefs of the feafon. If the ftock
be a laft year's fwarm, fet a duplet *over*
it; and as foon as that feems, by its weight,
to be three parts full, fet a triplet *over* the
duplet; which *laft*, when full, or nearly fo,
is to be taken off, and probably will be all
intire virgin honey, and without brood.
Then

Then raife the duplet, or double hive, by placing a triplet *under* it. But if the ftrength of the ftock is great, and there is plenty of honey pafturage, fo that another triplet may be expected to be filled, place the triplet over, inftead of that which was taken off. Perhaps, in fome good feafons and fituations, *three* or *four triplets* may be taken, if they are opportunely applied.

But if the ftock is of *two* years ftanding, it muft be raifed on a nadir; and as often as it requires enlargement take the fuperior hive off, and put a triplet in its place; and proceed thus as occafion may require.——— Thefe *two* methods of fuperhiving the *laft* year's fwarm *one* year, and the *next* of *nadir-*hiving the fame ftock, will be a fure means of obtaining the greateft quantity of *virgin honey*, and the largeft quantity of the *beft wax*.

Obferve, in all cafes, when hives are fet over another, that if the nadir is judged to be about three parts full, the door of it muft be ftopped, and that of the duplet opened, or the bees will not fo foon be tempted to afcend, to work in the duplet, nor will this

procedure

procedure increafe the labour of the bees in the meanwhile, as the way down is as fhort as the way up.

On the contrary, when a hive is placed under, the *door* of it muft be ftopped for a week or two, or till there is reafon to think there are fome combs made in it ; and then it is to be opened, and in two or three days after *fhut* again, difguifing it with a cloth, &c. hung before it, for two or three days.

Be particularly careful *not* to let the ftocks be *crowded*, before they are ftorified. For if a princefs is impregnated *early*, it may occafion a fwarm to rife fuddenly : for often great numbers of brood are hatched together, and therefore from want of room become ferocious, and occafion much inconveniency to the apiator and bees ; but prefently become peaceful and fatisfied on enlargement. For an additional hive having communications in direct lines with the combs of the hives added, the bees are led to efteem the whole as one hive, in a few days after its application.

In fome *critical* days or weeks, when *honey dews* are plentiful, or white clover or

F other

other pasturage is abundant, the quantity of honey collected in a few days will be almost incredible, if they have room enough to lodge it, filling a hive in *seven* days: often more than can be accumulated in a whole season.

But the advantages arising from additional hives are entirely lost in the old *single* method.

The duplets are in general not to be taken off till late, lest the queen should be therein, or it be mostly filled with brood. But super-triplets may be always taken as soon as filled.

Bees never begin to work in an additional hive, until new combs are wanted for eggs, or honey; and then the bees will begin to hang down, in *ranges*, or curtains, which is always a sign they have begun to make combs.

Bees often want enlargement before swarm time; which is denoted by their idly playing about the door and hive. It is the owner's fault and loss if he suffers it to continue.

Duplicated boxes will sometimes appear *full*
of

of combs and bees, through the back windows, though perhaps they are not above
a quarter or half filled, the combs being only
at the back.

If the bees of a triplet lie out, before the
ufual time of deprivation, it fhould be taken
and placed at a confiderable diftance, and the
duplified ftock raifed on a nadir hive: if, in
two or three hours after, the bees of the
ftock feem quiet, and work as before, as well
as thofe removed, it is a fign they have a
queen in each; and the hive taken may be
referved as a ftock, if fuch is wanted, or
fumed, and the queen taken away: moft
likely there will be much brood, which
may be fet over a weak ftock, or returned
again to its mother ftock.

In cafe duplets have idlers, they are to be
raifed on a triplet, and in about a month
the fuperior hive is to be taken off. For when
lying out in hot weather, though their hives
are not full, and the fwarming feafon is
paft, the bees will not *enter* notwithftanding;
but by adding a *nadir* hive, the accommodation of a fpacious and cool hall to regale

themfelves

themfelves will induce the idlers to enter it.

If it is fufpected that bees are idle (which, though they do not clufter out, may be difco-vered by their not being fo active as their neighbours), turn the hive up in the middle of the day : and if the combs are partly empty, it may be concluded they have either loft their queen, or fhe is unprolific, or is without drones; in which cafe they are to be flightly fumed in the evening, and fet over another ftock ; particularly a weak one to ftrengthen them.

But if the ftock is abundant in bees, and moft likely in honey, let them ftand till a young queen can be taken from a fwarm ; when placing her juft within the door, fhe will be joyfully received. Otherwife, if it is about the middle of the feafon, fume, and place them *over* a ftock ; and by that means it will produce a very large quantity of ho-ney.

Scanty breeders produce but little honey or brood; fo that, whilft other ftocks are rapidly increafing in riches, thefe will barely get enough

enough to fupport themfelves in the winter.

Empty combs placed in a duplet will not entice them the fooner to work therein ; for till the hive is completely full, and they are in want of others, they will not afcend, which in bad feafons may not happen for a confiderable time : neverthelefs, from being ready, they may be of confiderable advantage.

About the tenth of July the *upper doors* of all ftoried ftocks fhould be clofed, to induce the queen with more certainty to defcend, and breed in the *lower* hive, except it is *defigned to be taken ;* for then the door is to be fhut, and the upper one opened.

It often happens that in *poor fituations*, or in a long feafon of very *inclement weather*, neither duplets nor triplets will have work therein ; and this is *not imputable to a bad method* of management, or want of conduct, but *wholly* to a failure of the refources of pafturage, or of opportunities to gather it ; which fometimes has been fo great as to prevent the generality of ftocks from procuring a fufficiency for their own winter's

fupply.

supply. It is neceffary in fummer, when a hive has few bees, to ftrengthen it with a portion of bees from one that is ftrong. This will enable the queen to breed faft, and the hive will prove as profperous as any hive you have. But in all fuch reinforcements, the hive fo replenifhed fhould be fet at as great a diftance as your convenience will allow, for feveral weeks. This is a rule to be obferved in all fuch cafes.

Stocks that have *emitted* fwarms can but rarely be expected to yield a duplet that fummer, *unlefs* the fwarm is *returned*. Much lefs can a fwarm do it, though I have known fome exceptions in extraordinary fituations.

To *replenifh* a ftock that is *fcanty* of bees, fet fome empty combs, and pour the cells of one fide full of fugared ale, or platters of it, flightly covering it with a little hay or herbs, to prevent the bees from damaging themfelves in it : fet it on a hive floor in the morning, and place an empty hive over it, in the midft of the apiary.

A great multitude of bees will be attracted by the odour, and affemble round the feaft.

As

As foon as that is perceived, ftop the door of the hive until night; when the bees having afcended to the top of the hive, take it, and give them a flight fuming, and place them over or under the ftock that moft wants their affiftance.

If a queen is killed or dies in the fummer, it may be known by the bees not carrying in any farina, or by the door of the queenlefs ftock being much crowded, as well as that to which they carry the honey. Both hives appear prodigioufly active, as though a honey dew had commenced, and with a clear uninterrupted buz, with crumbs of wax about the door. Immediately ftop the door of the unfortunate ftock, and unftop it in the evening: the interlopers will then fly home. Early in the morning, take the hive to a proper diftance, and fume it; or keep them confined till next day, in a darkened room. They will then very peaceably and readily quit the hive on a little drumming on the fides. If the hive has much honey, cut the combs out; but take care of thofe that have brood, and add them to fome other ftock. The bees, however, will continue working till all the young are fealed up.

If

If a like accident happen in winter, take the bees out, put them to a ſtock, and take the honey.

In the want of a hive upon a ſudden demand of enlargement, and not having a proper one in readineſs, ſet a common one with bars acroſs it, in a pail or bucket, and place the ſtock over it; next night cloſe the joining, and at the accuſtomed time ſeparate it by the dividers, and take the bottom one away.

SUMMERS have ſometimes been ſo HOT as to ſoften the combs ſo much as to tumble them down, occaſion the ſmothering of the bees, and ruin of the ſtock. To prevent this, in ſuch weather, give them enlargement, and raiſe ſingle hives behind: ſcreen them as much as poſſible from the ſun, by large boughs, pouring often plenty of water about their hives, and taking off the hackels. Bee houſes ſhould have all their doors ſet open.

I   CHAP.

CHAP. XI.

THE NATURE OF SWARMS.

DURING the *winter*, ſtocks that are populous in the ſummer become reduced by
age and accidents to the ſmall quantity of a
quart, and the weaker ſtocks ſuſtain a proportional diminution. The *repeopling* the hives,
therefore, depends on the amazing FECUN-
DITY OF THE QUEEN, which furniſhes
thoſe new-born multitudes that conſtitute
the ſwarms.

In conſequence of a continued great increaſe, the bees feel a natural impulſe to
ſwarm. This law they are impatient to
obey, in defiance of all the obſtacles that
the ingenuity of man has contrived to its
taking place. A ſwarm does not conſiſt of
all *young bees*, but of *old and young* promiſcu-
ouſly.

The *breeding* of young bees is begun
ſooner or later, in proportion to the *fruitful-
neſs*

nefs of the queen, the populoufnefs of the ftock, the goodnefs of the fituation, and of the weather. The more numerous the bees are in the hive, the greater will be the heat to enable the queen to begin breeding *earlier* than thofe of other ftocks. When bees are carefully fupplied with food in fpring, they breed faft even in bad weather.

When *January* proves mild, the breeding will fometimes commence at the latter end of that month; but *often* in February, and in March *generally*. As foon as bees carry in farina, or yellow balls, on their legs, it is a fure fign of the queen's having begun to breed. A long feafon of cold and wet weather retards the hatching or increafing of the breed, caufing many abortions, and not uncommonly that of the royal nymphs. They may be feen caft out in fuch unkindly feafons.

The influence of a genial fpring haftens the breeding, and no lefs accelerates the bloffoms proper for their nourifhment; the fallows, willows, fnow-drops, crocufes, &c. yielding plenty of farina.

But fhould the weather be unfavourable

while

while thefe flowers are in bloom, thereby
preventing the bees from iffuing out to collect it, thofe already hatched will be ftarved;
and it will alfo delay a farther increafe,
until a more aufpicious change takes place.

If a fpring is not *very* cold, but wet, it will
not favour the production of royal brood;
yet the *common cells* will be filled with *young*,
but no addition of honey; which will caufe
the bees to be very anxious to fwarm, and
very irritable, flying about the hive in con-
fufion and difcontent. I have feveral times
feen *royal cells* in which the workers were
continually introducing their heads, I fuppofe, to feed the maggot; but, after a few
days, they entirely neglected them, probably
as being abortive. In fuch cafes no fwarm
can rife until another birth yields a princefs.

In fpring, when bees that are in no want
of food fuddenly give over carrying, it may
denote the unprolificnefs of the queen; and
if the hive contain but few bees, they had
better be united to another ftock.

In *forward fprings*, when the workers
are few, but the queen very pregnant, fhe
will be obliged to depofit her eggs fafter

4   than

than the fmall number of bees can fupply the maggots with fuftenance; and they will therefore perifh, and be caft out. This is a difadvantage which arifes from keeping *weak* ftocks.

To judge of the fulnefs of a hive in May, obferve the numbers of bees that enter the refpective hives, and form an eftimate.

Queens are not EQUALLY FRUITFUL. While fome breed flowly or not at all, others will fpeedily increafe in prodigious numbers. Sterile queens fhould be exchanged for the fpare queen of a fwarm; or at taking up time deftroyed, and a new ftock fubftituted.

From the middle of *May* to the middle of *June* is the moft *advantageous* time for fwarming; but they often rife, not only at the beginning of *April*, or fooner, but alfo as late as the 20th of *Auguft*; counties and feafons being fo very various. *Very early* ones are feldom large enough to conftitute a good ftock; and are in danger of perifhing if bad weather fucceeds. *Very late* ones, though moftly large, will often not have fufficient time to lay up an adequate

quate ftore for the winter, nor rear a brood in time : befide which, their emigration diminifhes the *parent ftock* fo much as to endanger its being ftarved during the next fpring. The prevention is, to encourage *timely* fwarms by *warmth*, and by a trough of fugared ale now and then, in February and March. But whether the fwarms are early or late, is a matter of no confequence in the ftory method, by which they are *returned* to the ftocks.

In a *good feafon* for early honey-gathering, the ftocks will not be forward to fwarm, though they have a princefs ready; being *then* wholly intent to collect the precious fweets, and almoft deferting the hive: the few left, finding fuch fpacious room, and full employment, have no temptation to rife, and quit fuch treafure for an empty hive.

Though a fpring fhould be *cold*, and *otherwife* unfavourable, a fwarm may *rife* the firft or fecond fine funny day, if a princefs is impregnated, notwithftanding the hive may be very *thin* of bees. The fwarm, of courfe, will be fmall. New fwarms will gradually defert their hive on a continuance of bad weather,

weather, and unite with another ſtock or ſtocks, without loſs to their maſter.

The *increaſe* of ſwarms in calm ſituations is frequently three from a hive; and ſwarms will emit ſwarms, or maiden ones. But it is to be obſerved, that in theſe caſes the *production of honey* is proportionally *leſs*, not near ſo much as might be expected from the multitude of bees, for the reaſons before aſſigned.

Frequently, when ſtocks *in very good ſituations* have many princeſſes, ſwarms will riſe though the weather has been unfavourable; while ſtocks only two miles diſtant may be ſtarving, and afford no ſwarm.

Stocks ſingle-hived, on being filled, and having a ſuitable princeſs, will often ſwarm repeatedly, though of ſmall bulk; by reaſon that, having no more ſpace to work in, they would rather ſwarm than be idle, that the precious advantage of honey-gathering may not be loſt.

In very DRY SEASONS few ſwarms are diſcharged. On examining the hives, no princeſs or royal cell was found. The cauſe is uncertain; perhaps the drought did not

favour

favour that kind of prolific nutriment fit to produce royal eggs or brood, and therefore no fwarms could be formed. Such ftocks fhould be taken at the feafon; for having, it is moft likely, none but old queens, they will die in the winter, and put an end to the ftock.

Bees that are placed near WOODS find therein abundant farina (the great fource of early fwarms) to feed their young. In all fituations that have plenty of farina, the bees are remarkably forward and active. In the HEATH countries, on the contrary, they are later in their productions than in other fituations, feldom fwarming till the end of July, owing to heath blowing late. In general, the bleaker the fituation the later the fwarms.

A WET EARLY feafon prevents the gathering of farina: then late fwarms will be the confequence ; and if the weather fhould continue very indifferent, they will rife when leaft expected, and be loft for want of watching.

After the *firft* or *prime* fwarms have rifen, the fucceeding ones fhould be *returned* to the ftock ; for if a fecond is emitted, it certainly fo much impoverifhes the ftock that little honey

honey can be collected afterwards, and will
not leave a fufficiency of bees to rear the
young, which at that time are abundant.
Undoubtedly there are exceptions, which a
difcreet apiator muft be left to judge of.
When *additional flocks* are *not* wanted, the
prime fwarms are to be *returned*, as well as
cafts; as being the moft *profitable* method.
The ftocks on the ftoried plan cannot be
kept too full of bees in the fummer time.

Stocks that have not fwarmed before the
firft of *July* from fingle hives, fhould be returned; but reference in thefe and the like
cafes muft always be had to the *difference* of
feafons and fituations, in which local circumftances only can direct the determination.

A large EARLY SWARM, with good weather fucceeding, will be far more productive
than a fimilar one that rifes later; for having
more time before them, their hives will be
furnifhed with combs and brood before the
honey harveft commences, and then are
prepared with empty cells and young workers, that will, in a fhort time, enable them to
collect a large ftore of honey, if care has
been previoufly taken to provide them with
fpacious

fpacious room. If bad weather fhould intervene, it will be prudent to feed them, for
which their fubfequent labour will amply
recompenfe.

There have been *inftances of ftocks* which
have fwarmed, and notwithftanding in the
middle of *July cluftered out*; and on having
another hive fet *over* them, ftill remained,
without afcending; but two or three days
after, on fetting a hive *under*, they prefently
entered, and worked vigoroufly.

With refpect to thofe ftocks which do
not feem to INCREASE in numbers, or appear to have DRONES; a dozen or two
fhould be taken from another ftock that has
plenty, and put to them. To effect this, in
a fine funny afternoon, when the drones iffue
out moft, take them fingly with the finger
and thumb as they pafs on the refting board,
and put them into a long phial, held ready
in the other hand, till the number wanted is
obtained: ftop the phial with a notched
cork, and at night faften the mouth of the
phial to the door-way of the hive, and by
morning they will have entered.

Thofe perfons who KILL THE DRONES

G in

in the *fpring*, are not aware that thereby
they are deftroying the only means of increafe; for the drones are to bees what males
are to other creatures.   But if it fhould be
obferved that the drones in *fummer* are fo
abundant, efpecially of a weak ftock, as
nearly to confume the honey as faft as gathered; in this cafe, and this only, fome of
them may be deftroyed.

Many *fchemes* have been tried for dimi-
nifhing the drones, but with little fuccefs.
For if they are ftopped from entering their
own hive, they know their next neighbours
will gladly receive them : nor will TRAPS
fufficiently deftroy them; and their application greatly difturbs and hinders the workers of not only their own hive, but alfo of
others.

More may be killed on the alighting board
in a fhort time by the end of a cafe knife,
than by any other means; and if done leifurely, the workers will not refent it for a
while. When they do, retreat, and try again
fome little time after. If continued long, the
workers will be fo difturbed as to *enter* other
hives, and the whole apiary be alarmed. If
the

the workers do not kill the drones at the usual time, a few may be killed by the fingers; and then thrusting a small twig into the hive will provoke the workers to finish the business. August is the usual time of massacre. The strongest stocks expel them the soonest: however, if they are not killed *then*, the cold weather effectually destroys them.

Great numbers of workers are bred before princesses or drones, which last do not usually appear before May, unless in early springs; and the populous stocks will have them in March, and often in April.

Hives, however, will often be so full of bees as to cluster out, and frequently swarm, without any appearance of *drones*; though it is probable there *may be a few*, but the weather too unfavourable for their shewing themselves, as they are more tender than the workers.

LYING or CLUSTERING.

THE lying or clustering out of bees, on or about a hive, has been commonly looked

upon

upon as a *fign* of their being ready to fwarm:
but this is deceitful. It indeed may denote
that there are bees enough to compofe a
fwarm; but it is alfo a token that there is *no
princefs* to go with them; for, in want of
room, they often continue cluftered feveral
weeks.

It muft be confidered, that when the combs
of a hive are *full of honey and brood*, the
fpaces left between, being only half an inch
in width each, contain only a *third* part of
the capacity of the whole hive—about four-
teen thoufand to a half bufhel—and confequently become *foon overcharged* by a forward queen, and the furplus is obliged to
lie out; which, in fact, they always do, in
fuch circumftances, and perhaps till the middle of Auguft in hot and dry feafons, when
but few bees can remain in the hive.

This cluftering is very prejudicial, not
only in the lofs of time, but alfo in *what* the
bees might have acquired by their labour
in that interval, ufually the moft productive
of any part of the feafon, when every bee
ought to be fully employed. Nor is this
all: the bees by this *indulgence* contract a
habit

habit of indolence not eaſily relinquiſhed. The example tempts others to be as idle as themſelves, greatly obſtructing thoſe that work, in their progreſs. Some, indeed, will be induſtrious in ſpite of their owner's inattention, and proceed to build combs on the outſide, or under the floor of the hive.

Although it is a certain ſign, when bees lie out from day to day, that there is no princeſs ready; yet as there is no practical means of knowing *when there will*, a conſtant watching is neceſſary.

Theſe diſadvantages are *admirably remedied by ſtorifying.*

But COTTAGERS, who have not this convenience, may cut a door-way in the back of an empty hive that already has one in front. Set the empty hive with one of its door-ways againſt that of the ſtock, fixing on a proper ſupport, ſo as to be on an exact level with the ſtock. The *vacancy* left between the two hives fill up with a piece of hay-band, &c. taking care, however, to leave the paſſage of the two door-ways *free.* The bees will then paſs through the empty hive

to

to the full one, till more room is wanted, and
then they will begin in the additional one.

To SEPARATE them when full, at night
gently take away the hay-band; have a lump
of clay or cow-dung of a proper confiftence
ready; nimbly force that between the two
hives fo effectually that it may ftop both
doors; take away the foremoft, and place
another empty one in its ftead the *next* night.
About an hour after taking up the firft, you
may venture to open the door of the ftock
with the end of a long ftick, and in the
morning entirely clear the dung away.

Under this management the bees will
conftantly be employed, nor can they poffibly be prejudicial to the owners, though
thereby the ftocks fhould not fwarm, for
doubling does not *prevent* it.

The PROFIT on bees depends, in a great
meafure, on the *detention* of the fwarms. If
THEY ARE LOST, the increafe of honey can
be but *trifling*, however carefully all other
particulars are obferved. A *cafual infpection*
will not anfwer this important purpofe. I
have not feen or heard of any apiators (my-
self

felf *not* excepted) who through neglect in this point have not loft, more or lefs, fwarms every year, and chiefly *prime ones*; for bees often fwarm without a minute's notice, perhaps the very inftant after being left. There is no fure way of fecuring fwarms but by a CONSTANT WATCHING of a *bee-herd*, retained on purpofe, from feven or eight iu the morning until three or four in the afternoon, till all the prime fwarms have iffued. Bad weather may be excepted.

Children, or rather aged people, might be employed to do it at an eafy rate; and if it fhould coft feven or eight fhillings, it is better to be at that charge, than run the great rifque of lofing feveral of the *beft fwarms*. You alfo efcape the anxiety and trouble of going conftantly to and fro, which is after all attended with uncertainty. Befides, if a perfon keeps but fix ftocks, and faves only one fwarm, he will be no lofer; to which add, the affiftance given to an indigent family by the money expended. The ufual hours of fwarming are from ten to two; but this is not to be depended on. I

G 4 have

have often known, and had them *rife* as early
or late as the hours ftated.

Another caufe of the lofs of *prime* fwarms,
is the *miftaken notion* that bees always fhew
certain *figns* or *tokens* of their going to
fwarm; and therefore until thofe figns appear, watching is omitted. But it muft be evident to every reflecting apiator, that fwarms
frequently rife early in the fpring, as well as at
other times, without fhewing any fuch figns
at all. On the contrary, in fome feafons
the hive may be fo very full of bees as
largely to clufter out, and make an aftonifh-
ing noife within, as though that moment
they would rife, and yet very often do *not*;
no, not for *feveral* days or weeks afterwards,
and fometimes not at *all*. Thefe tokens,
indeed, clearly fhew there are bees fufficient
in number for a fwarm, and they are moft
anxious to do fo; but it alfo fhews they
cannot break nature's law: NO QUEEN NO
SWARM.

Although there are no figns that precede
firft fwarms; of *fecond*, or *cafts*, or after ones
there are, viz. peculiar SOUNDS or NOTES
in

In the hive not heard at any other feafon. They may be heard generally in the evening in fine weather, and fometimes for feveral days together; being probably expreffive of the princeffes' being ready, and defirous of enjoying empires of their own; for feveral are afterwards heard at a time, in a kind of refponfe either more acute or grave, but very different from any founds made by their wings, and feeming to be formed by a tube, refembling the expreffions of *toot, toot, toot,* or nearly that of a child's penny trumpet, but not near fo loud. Many chimerical conjectures have been formed relative to this particularity; but one *certain meaning* they convey to the apiator, that when heard he may be affured that the *firft,* or *prime fwarm, has efcaped,* if that will comfort him.

It indicates alfo, that a fwarm may be expected very foon, perhaps the next day, or in a few following ones, according to the finenefs of the weather.

When the number of *princeffes is too many* to be fupplied with bees for fwarms, it in-

duces

duces three or more to iffue with a fingle fwarm, and either fettle together, or divide into different clufters; well knowing that death will be the fate of thofe that tarry behind. Sometimes, indeed, a princefs will coax *a few* bees to accompany her, and form a fmall caft, of no profit, but which rather contributes to impoverifh the ftock.

SECOND SWARMS are feldom worth preferving *fingle*; but by uniting two or three, you may form a good ftock.

If a fwarm is wanted from a *duplet*, both doors muft be left open; but if none fhould rife, the ftock at feparation moft likely will have a queen in each.

It is very probable that a princefs may fometimes rife unimpregnated, or not ripe for *laying*, and which the bees at their exit with her were not fenfible of; but when hived, finding their miftake, they abandon her and the hive, and return home again.

On the *rifing* of fwarms, many bees juft returned from the fields with their loads, and many juft entering, join them; by which means they are capable of conftructing

ing

ing combs prefently after fettling; and fometimes do on the branch of a tree, if they are
fuffered to remain there a confiderable time.

When bees *play idly* about the door on
hive, and are more than ordinarily mifchievous; it is a fign they are anxious to
fwarm ; and probably may rife, though -
without a princefs, if it is late in the feafon,
but will return home again.

If the wind be brifk at the time of a
fwarm's rifing, it will fly in the fame direction, and will fettle in that fpot which will
beft fhelter them from the inconveniency,
regardlefs of their *accuftomed* place of cluftering.

As none but good fwarms at any time
ought to be kept, it will be neceffary to
afcertain how fuch may be known. It
fhould be in bulk, when hived, not lefs than
a *peck and a half;* in middling fituations
they run more. I have had them in Hertfordfhire frequently half a bufhel, fometimes larger.

Near Pembroke they feldom exceed a
peck, which is here efteemed a good fwarm.
How-

However, not *less* than a peck will prove a productive one.

A swarm will appear much larger as it hangs on a bush, than when clustered in the top of a hive.

The number, weight, and measure of bees.

		lb.	oz.	dr.	
100	drones	0	1	0	
290	workers	0	1	0	
4,640	——	1	0	0	AVOIRD. WEIGHT.
915	——	0	3	2	
1,830	—— a pint	0	6	5	WINCHESTER MEASURE.
3,660	—— a quart	0	12	10	
29,280	—— a peck	6	5	6	

This statement is made on an average; for they will not prove twice exactly alike, because of their different degrees of fulness, &c.

CHAP.

CHAP. XII.

THE HIVING OF SWARMS.

As swarms (pl. 2, fig. 2.) frequently rise when not expected, and that with precipitation, common prudence, it might be thought, would induce apiators to have hives in readiness. But I have often seen the contrary, though the expence of the hives would be less, when bought early, and you would also avoid the risk of losing a swarm while seeking a hive.

The poverty of cottagers may be an excuse for such supineness. Therefore in such an exigency the swarm may be put in a pail, bucket, basket, &c. in which let it remain till the evening; when turning the vessel up, lay two flat sticks across it, place on it an empty hive, bind a cloth round the juncture (all but the door-way), and by the morning the bees will have ascended therein; but if not, gently beating the sides of the vessel will cause them to ascend.

To

To PREPARE hives for the reception of fwarms, the fnags, or the roughnefs of the ftraw, fhould be clipped off, and rubbed as fmooth as can well be, as this will fave the bees a deal of labour, which they will employ to greater advantage in conftruding of combs.

Boxes fhould have all holes and crevices ftopped with putty, or other cement, which otherwife the bees muft do, to exclude air and vermin.

SPLEETS, or fticks, are proper to fupport the combs, when extended near the bottom; but TWO ONLY are neceffary, and placed thus +, at the height of the fecond round of ftraw from the bottom; one from the front to the back, the other acrofs *that*, from right to left: for as the combs are *ufually* built in parallel lines from front to back, each comb, when wrought down, being of confiderable weight, it will have a ready fupport from the fpleet, and which will ferve to faften them alfo; but till they become weighty, *no* faftening but that which the bees themfelves execute, will at all be needful.

But

But as *fometimes* the combs are conſtructed obliquely, or tranſverſely, a ſecond ſpleet
is neceſſary to take them in that direction.
In fact, *common hives*, having no occaſion for
removes till they are taken up, *need no ſpleets*,
as verified by bees in hollow trees, &c.
However, the two mentioned are enough for
any hive, even in the ſtory method : much
leſs have they occaſion for *any* ſpleet near
the top, and which is generally ſo prepoſ-
terouſly placed as to be very troubleſome
and prejudicial to the honey, in taking the
combs out.

No other *preparation* or *dreſſing* of hives
is neceſſary, than that which I have mentioned. The employing herbs, and many
other fanciful articles, *is of no uſe;* but as
people are wedded to old cuſtoms without
rational foundation, ſugared or honeyed ale,
ſprinkled in the top of the hive, is the moſt
alluring ſubſtance that I know of. The
truth is, when a ſwarm quits a clean hive,
it is for *other* cauſes, and not through diſtaſte
of the hive, unleſs it is too ſmall.

It is cuſtomary to make a TINKLING
NOISE to ALLURE ſwarms to ſettle. Why
it

it does so is *uncertain*, but that it does is as certain. Most prime swarms, that are not in a habit of settling in an usual spot, are mostly lost, if not *tinkled*.

Besides which, it ascertains the right that the apiator who follows it, has to claim it, if strayed from his own premises. The greater the noise, the sooner it is likely to succeed. I find a WATCH RATTLE (used about London) the most efficacious, and that when the common method has failed.

In *prime* or first swarms, the noise should not begin till such a quantity of bees have arisen as will form a good swarm, for fear of terrifying the princess from issuing; and if so, all the bees will return, though hived. A sudden storm, dark clouds, or thunder, will cause them to return, if not settled; or if the princess, too weak to sustain the flight, drops on the ground; or if the bees are *roughly* treated in the hiving.

The noise should be made on the contrary side to that which will be most proper for settling. Nor should it continue longer than the bees *begin* to cluster: there is no danger but the rest will follow on hearing their buz.

buz. When they rife in windy weather they are very irritable, and apt to fting; and though cluftered, often return home.

When a prime fwarm is broke or divided, the *fecond* will be much fuperior; and therefore, if it is in good time, may be kept, if a ftock is wanted.

When bees are hived, but feem difcontented and tumultuous, it is a fign they have no queen among them. Probably fhe will be found on the ground, with a fmall clufter furrounding her. Take the clufter up, and place it on the outfide of the hive which has the fwarm, or near the door; it will foon make them eafy, and allure thofe on the wing to join them alfo.

Hives fixed near the fpots where bees have been ufed to fettle, and rubbed with fugared ale, will *fometimes* decoy fwarms to refide therein. But this muft *not* be relied on; for it often happens that bees previoufly choofe a place, that they have made clean for their reception, and to which, on rifing, they immediately repair. But a hive of old combs will certainly allure fome of

H   your

your own fwarms to fettle therein, if not of
fome ftrayed ones.

If a fwarm is too *large* to be contained in
a hive, immediately double it; but if it is a
common hive turn it upfide down in a bucket, &c. and lay two flat fticks acrofs, and
fet another hive over it; then take them
from the bucket, and fet them on four or
five rounds of an old ftraw hive *doubled,*
as they are, and in the evening place them
on their deftined ftation, ftopping the joining with clay, and allowing a proper door-
way.

When fwarms feem reftlefs fome time
after hiving, as often happens from their having two princeffes, and being undetermined
in their choice; take them to a dark apartment, when the bees, fuppofing it near night,
will prefently elect the moft promifing
lady, and expel the other.

It is very likely that the old queen fometimes accompanies the firft fwarm. The
reafon perhaps of there being *no tooting preceding the firft* fwarms, is there being then
but one young queen qualified to lead them.
                                        When

When more *royal cells* than one are perceived in a hive, the fupernumerary ones may be taken out to make a fwarm, if wanted.

Bees, when fwarming, are generally very peaceable, as being under many fears and apprehenfions; fo that they may be hived with much eafe and little danger (unlefs the wind is high), if they are treated with gentlenefs.

If they feem inclined to rove beyond the proper bounds, handfuls of fand, dirt, or the like, fhould be thrown up among them: water alfo caft among them will induce a fpeedy cluftering. The fame *means* fhould be ufed when two fwarms rife together, and fight in the air. A great noife fhould be made, efpecially that of a gun, to intimidate them.

If feveral PRINCESSES rife with *one* fwarm, when hived together, great commotions enfue, until one of the princeffes is caft out or killed.

But when they cannot decide in their choice, they fly out, and continue the conteft; or, which is moft frequently the cafe, different parties clufter with the lady they

approve, and fettle feparately. Let them
alone till they are feverally fettled, hive
each parcel feparate, afterwards ftrike them
out on a board one after the other, and take
the queens from each, all but the largeft
clufter, to which put all the reft. Or otherwife, at the clofe of the evening, *fume* them
all together, when the firft princefs that recovers will be acknowledged queen, and the
reft expelled or flain by the morning.

STRAY SWARMS are often perceived flying in the air, and may be allured to fettle
(efpecially if tired with flight) by making
fome kind of tinkling with a knife upon a
fork, fhovel, or the like; and when fettled,
may be brufhed into a hat, handkerchief, or
part of the garment, which being gathered
up by the corners, may fafely be carried
home, and laid on the ground, or table;
laying a ftick acrofs; and placing a hive
over them, they will affemble therein.

When a *fwarm* fettles in SEVERAL CLUS-
TERS, hive only the *largeft clufter*, and re-
move it, a fmall diftance at a time, near to
the fmaller clufters, which are fucceffively
to be fhook off the places of cluftering by a

9   long

long hooked ftick, repeatedly, till the buzzing of thofe in the hive has attracted their
notice, and induced them to join. If the
clufters are equal in bulk, hive both feparately, and fet them at a fmall diftance from
each other; and if either of them have a
queen, and are diffatisfied with her, they
will quit the hive, and unite with the other;
but if both remain contented, unite them by
fuming.

Swarms fhould be hived as *foon* as fettled : for their cluftering is generally but of
fhort duration ; efpecially of prime fwarms,
or if they have previoufly felected a place of
refidence.

When a fwarm attempts to *fettle on a
perfon*, ftanding or walking, &c. let him not
be alarmed, nor in any wife oppofe them, but
lift the hat a little above the head ; perhaps
they will fettle on that : if not, cover your
head and face with a handkerchief for them
to clufter on. But if, contrary-wife, they
begin to clufter on the fhoulders, or under
the handkerchief, fling it off, and fpread your
hands over the eyes and face, and thus re-

H 3 main

main *entirely paffive*, till the whole have fixed,
which, if this is punctually obferved, will be
done without a fingle fting. Then retreat
with leifure to fome room in a houfe,
made nearly dark, and then a perfon muft
hold a hive, pan, fieve, &c. (fprinkled with
fugared ale) over the clufter, with the edge
juft touching it, which will, after a little
while, induce them to afcend into it. Blowing with *bellows* will caufe them to do it the
fooner, without irritating their propenfity
to fting. But if any violent or offenfive
means are ufed, it will provoke their revenge
fo as to be dangerous.

When a fwarm is cluftering, and ANO-
THER is rifing and endeavours to join it,
cover the firft with a thin cloth, and throw
duft, or water, among the others, to caufe
them to fettle elfewhere. As likewife if
a *fwarm that is rifen* attempts to fettle on a
ftock hive, ftop the door, and cover the hive
with a cloth. Sprinkle an empty hive with
fugared ale, and place it a little raifed over
the top of the ftock, and the fwarm will
enter therein. If the fwarm feems too large

to

to be contained in the hive, fet another upon the firſt. As ſoon as the bees have entered, take it away, and unſtop the ſtock.

OR it may be done by ſtopping the door of the ſtock, and immediately removing it to ſome diſtance. In the interim an aſſiſtant is to place an empty hive in its place, to which the ſwarm will enter; and then it is to be taken to an appropriate ſtand, and the ſtock brought back to its former ſituation.

SWARMS will ſometimes cluſter on, or enter, *improper places*, as under roofs, or other buildings. Immediately a hive is to be placed cloſe by, or juſt about the hole of their entrance: encompaſs the hive and bees with a cloth, and it is very likely after a little time they will give the preference to the hive. If not, put a piece of paper with holes made in it over the bowl of a pipe of *tobacco*; apply the end to a ſmall hole made juſt under where the bees entered; take the empty hive away, and then blowing forcibly, the ſmoke will generally induce them to fly out, and cauſe them to ſettle in a more convenient ſituation for hiving.

To

To avoid repetitions, I would obferve, that the GENERAL RULE in conducting operations about bees is, that they be executed without noife or talking in approaching the hives, till the doors are *fecured*; otherwife the bees will be alarmed, and guard the doors immediately. A leifurely and calm deportment, with gentlenefs yet boldnefs, and giving the leaft difturbance, will greatly conduce to render the bufinefs eafy and fafe.

In HIVING take care that none are *crufhed*, as that provokes the others to revenge; and not only fo, but it may chance to be the queen, to the ruin of the fwarm. Forbear the ufe of weeds, or throwing water on them, when cluftering, or brufhing them off, which they will highly refent; and it may make them fly quite away. Gently cut away all fpray twigs, or branches, that may obftruct the placing the hive under the clufter. Always *fpread a cloth on the ground*, with two fmall wedges on it, as near the clufter as may be: the wedges are to keep the edges of the front of the hive a little raifed, for the more ready entrance of the

the bees underneath; as alſo to prevent in-juring any of them.

It may be remarked that ſwarms often ſettle *without a queen*; which, therefore, proves, that it is not the queen that leads and begins the cluſter. Moſt likely thoſe that are moſt inclined ſettle firſt, and the reſt naturally follow; as ſheep through a hedge.

INSTRUMENTS neceſſary for hiving are, an empty *box* or *hive*, a hive floor, or looſe board, a large cloth, two ſmall wedges, and a long fork, or crook-ſtick.

To HIVE BEES, let the apiator take the hive inverted, and leiſurely introduce the hive under the cluſter as conveniently as can be without diſturbing the bees; then with the left hand give the bough two or three ſmart ſhakes, which will cauſe the greater part of the cluſter to fall into the hive: *nimbly* take it away, and turn it on one edge on the floor, and the other on the wedges; draw the cloth up over the hive, leaving the raiſed part open. The bees, as may be expected, will be in great confuſion, and make a great buz, but will immediately

begin

begin to afcend : the bough, or bufh, &c,
muft continually be fhook by the long ftick,
whilft any bees endeavour to relodge on
it : thofe on the wing, hearing the buz of
their companions in the hive, will gradually
fly down and join them.   Let them remain
on the fpot till the evening, unlefs the fun
fhould be too violent; and then the heat
would make them *quit* the hive, unlefs fhel-
tered by boughs, or the like. But if it fhould
be inconvenient for the hive to remain, they
may be removed a little way off. As foon as
the bees are nearly retired into the hive, the
hive may be carried to its deftined ftand ; the
few bees that remain on the wing will return home.

Whenever bees are fo cluftered that a hive
cannot be put under them, lay a cloth under, or as near as circumftances will allow ;
fhake the bufh, &c. to make the bees
fall, and keep fo doing till the bees relinquifh it : when down on the cloth, or
ground, fet a hive over them, and they will
enter.

Or, fhould a fwarm fettle on a hedge, &c.
that a hive cannot be fet under them, it may

be

be placed OVER them: this do by forked
ftakes, or cords; and by flinging a cloth over
the bees and empty hive, they will in
fome hours afcend. But for fear of mifchance, they fhould be *watched*. Or if they
are found not to afcend, fet the hive three
parts over a floor, then with a *fpoon* very
tenderly take up fome of the bees, and turn
them out on the floor, within, or near the
door of the hive (its edge being raifed by a
wedge): repeat it as long as the bees will
permit without fhowing much refentment:
the buz of thofe already entered (the larger
the number the better) will the fooner allure the others to do fo. But if the bees are
fractious at firft, introduce only a fpoonful
or two at a time; and in the intervals retire
out of fight.

Or to prevent a fwarm from CLUSTERING
INCONVENIENTLY in a hedge or bufh,
immediately lay a handkerchief or hat on
the bufh: probably they may fettle on that,
and may afterwards be laid on the ground;
and a hive being placed over, they will moft
likely embrace the offer.

BEES cluftering round the BODY OF A
TREE,

TREE, OR POST, are difficult to hive. Take a *hive* and *floor*, or board, and place it by means of forked sticks, barrels, ladders, &c. or with cords, so that the floor may be on a level with the bottom of the cluster: then raising the edge of the hive next to the bees, by wedges, gently advance the hive so as slightly to touch the cluster: this in a little while may induce some of the bees to enter, and the rest to follow. But to save time, use the *spoon*, as before directed, to diminish the cluster, and increase the buzzing in the hive: at times disturb the cluster, by gently shoving a small stick among the outermost, to disengage them. As soon as a considerable number have entered, the rest will surely follow; though, perhaps, but slowly; unless the queen has been one of those conveyed by the spoon.

Should swarms fix on the EXTREME BRANCHES or twigs of high trees, beyond the reach of the hand, a hive, or rather a light basket, must be suspended to the end of a long pole or fork. Then having a ladder, introduce the basket under the cluster, while an assistant with a long crook smartly

shakes

ſhakes the bough, by which a great part of the bees will fall into it. It muſt then ſpeedily be brought down, and turned upſide down on a cloth ready ſpread, on which many bees already fallen will be ſettled. In the mean while the branches muſt be conſtantly ſhook, by which the bees, finding no quiet there, and hearing the buz of thoſe underneath, will deſcend and join them.

Or, *another method* is to tie twigs to the end of a long pole, and therewith diſturb the cluſter till they take wing again ; when probably they will cluſter in another ſituation more favourable, if treated with the uſual muſic.

A third means is to hold a pan of ſmoking ſubſtances, which may make them glad to move their quarters.

When ſwarms ſettle on LARGE BRANCHES of trees, too ſtubborn to ſhake, a hive is to be ſet on a floor, and faſtened with cords, that the floor may touch the cluſter. Then treat them as before mentioned.

A ſwarm in a HOLLOW TREE that has not been lodged therein more than two or three days, may be diſplaced, by carefully ſtopping
ping

ping all the holes, and crevices, except that
which they entered by ; then fixing the bottom of a hive againft their hole of entrance,
fecuring it firmly with cords, as alfo tying
a cloth round the joinings, that no bees can
efcape ; beat with a large hammer, or great
ftone, violently about the tree juft below the
hive : probably this will terrify the bees, fo
as to induce them to feek fecurity in the
hive. Now and then ceafe the noife, and liften
whether they make a buz in the hive ; and
repeat the hammering until the buz is
greatly increafed. Then, loofing the hive
from the tree, fet it on a cloth fpread on the
ground, and repeat the ftrokes and noife on
the tree till but few bees rife. Stop the hole
of the tree, and thofe on the wing will rejoin their companions.

But if they will not *take to the hive*, make
a hole with a chiffel, near the upper part of
the hollow (for the bees generally lie as
high as poffible above the entrance) : place
the hive juft above the hole cut, and by
hammering it will caufe them to fly furioufly
out, and take to the hive, or fettle in a more
commodious fituation. But if they fhould
have

have fettled below the paffage hole, make the large hole *below* the clufter, as near as can be judged, by ftriking where the buz may direct.

If thefe methods prove unfuccefsful, re-courfe muft be had to *fmoking rags*, damp ftraw, or cow dung, put into the hole, if it be made large enough; and at the fame inftant hammering under their lodgement, or teafing them by thrufting twigs up till they fly out. Perhaps (for I have had no opportunity of trying) if an opening could be made large enough to receive a pot of fuming *puffs* under them, for about twenty minutes; by confining the fmoke, pro-bably the bees might be fo ftupefied as to fall to the bottom, and might carefully be taken out, by a ladle, or fpoon, and put into a hive, and immediately carried away, and placed in a dark room or out-houfe till the morning. The chafms of the tree fhould be all ftopped to prevent the bees from re-turning to their former lodge. The longer bees' have fettled in any place, the lefs dif-pofed they will be to quit it; efpecially if they have made combs, and have brood therein.

therein. They will fooner die than quit it. In fuch a cafe it is better to let them remain till autumn; and then fuffocate the bees and take their treafure.

Bees in the holes of walls may be treated after a fimilar method.

But when bees have fettled under the *roofs or vacant parts of buildings*, where fparks of fire might be dangerous, fuming muft be avoided; and inftead thereof WATER muft be conveyed over the bees, by the rofe of a watering pot, funnel, or pipe, taking fome tiles off, or boards down, to come at them; which will often fucceed as well.

Where WINDOWS have been left open, fwarms fometimes affume the liberty of taking poffeffion. To fecure them, firft fhut the window and door; then holding a hive under the clufter, draw a wire or thin ftick gradually between them and the cieling, or part to which they are attached: this will caufe the bees to fall into the hive; which being fet on the floor, the ftragglers will foon hear the buz of the others, and rejoin them, and the fooner if the room is made *nearly* dark.

All

All fwarms, if the weather is fine, will begin to work as foon as hived; but if the
two firft days prove foul, it difcourages
them from labouring for feveral days, even
if then it fhould be fine. But in a long
continuance of bad weather, they will *perifh*,
unlefs relieved by a timely *feeding*.

The foregoing *directions*, it is prefumed,
will be fully applicable to all other cafes
that may arife, though attended with fome
variation.

I   CHAP.

CHAP. XIII.

ARTIFICIAL SWARMING.

I AM forry to declare, that I have met with no *invention*, among the *many* that have been publifhed, or among the great number of my own devifing, for artificial fwarming, ADAPTED TO COMMON USE, or that has been in general fuccefsful. From fo great a difappointment, I am inclined to draw a conclufion, that as nature has implanted in bees a ftrong propenfity to fwarm, as a quality neceffarily connected with the manner and feafon; all our attempts, by *force* or *allurements*, to *effect* or prevent it, with a tolerable degree of timely advantage, muft prove ineffectual. I propofe the two following methods, however; as, if not fuccefsful, they will not be prejudicial to the ftocks, may amufe the *curious*, and be accomplifhed without much trouble. But they are *inapplicable to general* practice.

By

By often looking through the windows of ftoried boxes, in the fwarming feafon, SOME-TIMES a queen may be feen in one of the boxes. *Immediately* fhove a divider between the two boxes. Leave them about an hour; when if the bees of both boxes remain quiet, wait fome time longer, and then repeat the infpection, by intervals, two or three times, till the approach of night; and if they are ftill in a quiet ftate, introduce the other divider, and take the duplet to a diftant ftation. On the contrary, if the bees of *either* box have fhowed figns of difcontent, it is a token there is no queen in that which fhows uneafinefs; and therefore the divider muft be withdrawn, till another favourable opportunity offers.

The SECOND METHOD is: In the fwarming feafon, when the bees feem very numerous, and fhow indications of fwarming, fhove a divider between a duplet in the morning, having before opened both doors; and if the bees remain quiet and purfue their work, in both boxes, till the evening, proceed with them as above. But if the bees

I 2   of

of either box are confufed, take out the divider, and try your fortune another time.

An artificial fwarm may be made, by
purchafing one or more of fecond or third
fwarms of your neighbours, as they will be
of little value to them, and therefore may
be had cheap. Unite as many of them in one
hive, as are fufficient to form a good fwarm, by
placing the feweft in number to the moft populous ; fuming them firft to prevent quarrelling. But if fuch fhould happen, fumigate
the duplet.

CHAP.

CHAP. XIV.

OF WILD BEES IN WOODS.

IN February and March bees are very frequently numerous, on fallows, ofiers, and
other plants that afford farina, in WOODS;
which is a fure token that their habitations
are not far diftant. They may be eafily
traced; and having found them, mark the
place or tree. Aged people, or children,
may be fet to watch their fwarming, and
they may be hived in the ufual manner.
For whether in hollow trees, or any other
habitation, bees equally caft out fwarms, as
well as thofe in hives. Having fecured
and carried away the fwarms, in autumn repair to the fame fpot, and take the fummer's
produce, as directed under hiving.

If this early attention has been neglected,
make obfervation in woods on thofe places
which are moft plentiful of bee-flowers; or,
in very dry weather, of watering places, to

I 3   which,

which, in such seasons, they will be obliged
to resort. If their abode is too far to be
traced, dissolve some red or yellow oker in
water, and, dipping some sprigs therein,
sprinkle the bees therewith as they alight.
Being thus marked, they will be easily dis-
tinguished. For, by observing whether re-
turns are sooner or later, or whether in
greater or lesser numbers, a tolerable guess
may be made; especially after a little practice. A person having a watch, may by it
more accurately determine this point. A
pocket compass will also be greatly assisting
to certify their course, which is always in a
direct line to their habitation in their return
home.

If this method proves not successful, take
a joint of a large reed, or of kex; force a
part of the pith out at one end, and do the
like at the other, only leaving a small parti-
tion between the two hollows; cut a small slit
over one of the hollows, put some honey made
a little damp with ale in the hollow, and stop
the end with a cork, or paper; and if fire
can conveniently be had, melt some wax on
the tube, the smell of which will be wafted

by

by the wind to a great diftance. Place this joint near their haunts, and they will foon be allured to enter into the hollow. When about eight or ten have entered, ftop the end with the finger; foon after let one-of the bees out, purfue it as long as it is in fight, and then let out another. If it con-tinues the fame courfe, follow that alfo.; but if any take a different route, let another fly, and fo proceed till you find feveral take the fame courfe, which will lead to their nefts.

The bees that purfue other directions probably belong to other nefts, which may be difcovered by the fame procefs as the firft,

If it be neceffary to take the combs out directly, a pot of fuming *paffs* fhould be in-troduced under them by a hole made on pur-pofe. During the fumigation forcibly ftrike the tree. If the whole are not fallen from the combs, they will, however, be fo lethargic as to give the operator but little annoyance, if he has on the bee-drefs. The combs are to be taken out as whole as poffible, and placed in an empty hive, and fupported by as many fpleets as are neceffary, in the beft

I 4   manner

manner the nature of the cafe will admit of.
The ftupefied bees which have fallen into
the cavity of the tree, may be taken out by
a fpoon or ladle, and put to the combs in
the hive; which had beft be fet on a floor
before the combs are put in, and then the
bars and cover, and may be removed without much trouble or difplacement.

If puffs are not in readinefs, the fmoke of
dried cow dung, damp ftraw, &c. may be
ufed, which will be likely to force the bees
out; when, fettling on fome tree, &c.
they may be hived, and, on being carried
home, may be fet over the hive of combs.

If the nefts are taken during the fwarming feafon, thofe parts of the combs that have
honey in them may be cut out, taking
great care of thofe with brood, which, with
the empty ones, are to be placed in the hive,
as well as can be in the fame manner and
at the fame diftance as the bees do; and
placing the bees in them, they will foon repair the damage, and furnifh the hive
afrefh.

When the bees are efteemed not worth
preferving, rags dipped in melted brimftone,
and

and put under their nefts, will immediately fuffocate them.

Hives rubbed with honied ale, and fome poured into an old comb, and put under them, and placed on thofe fpots which bees much frequent, will be likely to allure fwarms to fettle therein.

Having had no experience in what relates to this article, the above is given from refpectable authority.

CHAP.

CHAP. XV.

SALVATION OF BEES.

MANY of my readers will be much surprifed at the following declaration, viz. That the SUFFOCATION of bees kept in *common hives* is *not* prejudicial to the intereft of the owners. This affertion, I beg leave to ftate, relates *only* to thofe who keep bees in SINGLE HIVES, WITHOUT STORIFYING.

Contrary to my former principles, prejudices, and practice, and to the current opinion of writers, nothing lefs than a feries of ftubborn facts could have effected my conviction and recantation.

From theoretic deductions, to facts I appeal;—to experiments, the juftnefs of which the judicious apiator may be convinced of, by making proper obfervations. For thofe who keep bees in boxes, with large windows, may perceive that in December and January very few bees are to be feen in the boxes that were crowded in Auguft. Thofe who

have

have ſtraw hives may, at that time, ſafely
turn them upon their edge, and have a tolerable view, to anſwer the above purpoſe.
The diminution is ſo great, that the fulleſt
hives or boxes are then reduced to about a
QUART! and this by the natural decreaſe of
the aged bees.   To certify this, I took the
bees from ſeveral hives, and found them to
meaſure as by the above ſtatement; the
weaker ſtocks leſs in proportion.

This reſult proves, that all the advantage
obtained by ſaving the bees of STOCKS
TAKEN, and uniting them to other ſtocks
(the only eligible means of ſaving), is ulti-
mately only the ſalvation of a quart.   And
as the queen muſt be killed by the hand, or
by the ſtock bees to which they are to be
united, they cannot poſſibly make any farther increaſe in the ſpring.

The queſtion is then reduced to this iſſue:
Whether the multitude of bees, united about
Auguſt, will not conſume (though gradually
diminiſhing) more honey before the ſpring
gathering commences, than the quart left
will compenſate by their labour?

Beſides,

Befides, it is to be confidered, that the eggs produced by the old queen of the ftock, not being more than ufual, want not an unufual number of workers to rear them; a greater number may poffibly be ufelefs, or prejudicial by the increafe of confumption. Nor do they contribute to the production of more early fwarms; for that depends on the early birth of princeffes, in which the additional bees have no fhare.

The truth of the fact is further confirmed by experiments on ftocks that have the bees of other hives united to them, but which proved neither more *forward* nor more *productive* than fingle ones hived in the common way, not only of my own, but of neighbours.

On the contrary, STORIED ftocks, in the fame feafon, were abundantly more profperous, having provided themfelves with means fufficient for their own profperity, in a fucceffion of peace and plenty, and without the cruel *neceffity* or *trouble* of *fuffocation* by fire and brimftone.

From

From this declaration it by no means follows, that the *old practice* of fuffocation can be juftified ; but muft be condemned as impolitic, and highly difadvantageous ; *for they muft be very weak who purfue a plan of conduct of fmall profit, when a better is offered of double or treble advantage.*

CHAP.

CHAP. XVI.

BEE GLASSES.

THE moſt convenient ſhape to ſet over bees, ſhould be ſimilar to thoſe of pl. 2. fig. 1 ; that is, perpendicular to the circular top, or ſtraight dome.

Four are deſigned for a box, *one* at each corner; and *one* in the middle which is to hold *two* quarts; the others, only *one* quart each.

That of the centre ſhould be in two parts; the lower part to be open at both ends; the upper diviſion of the glaſs to be circular at top. There muſt be a thin circular piece of wood, of proper dimenſions, to lay over the top of the under glaſs, to ſupport it when ſet over, and in it three apertures, cut out from the middle, by which the bees are to aſcend into the upper half of the glaſs.

It will be neceſſary to have an ADAPTER, or board of the ſize of the top of the box, on which the glaſſes are to be ſet. Apertures are

5

are to be made in it, to correfpond with thofe on the hive-top, but to be limited in length, and not to exceed the width of the glaffes, as pl. 2. fig. 3.

Inftead of fticks to fupport the empty combs, STAGES feem preferable. Three flips of wood, an inch and a half wide, and of a length to fuit the bottom of the glaffes: fmall holes are to be made near their edges, to receive long pegs, or flight fticks, about three or four inches long, and thus form *ftages* wherein to fix the empty combs. The bottom edges, and ends of the ftages muft be round, or bevelled off, and the ends of the pegs are to be cut fmooth with the furface, to prevent any impediment to the entrance of the divider. The fmall glaffes require two fuch ftages ; the larger central, *three*, in each divifion ; and to be placed fo as not to obftruct the apertures of the box by which the bees muft afcend into the glaffes.

OBSERVATIONS.

When the glaffes are filled with combs, the edges are to be cut through with a thin knife,

knife, clofe to the glafs; and a ftiff wire, bent like an L, with its fhort end made flat and fharp, is to be introduced between the combs. Give it a twift, to turn the flat end, fo as to feparate the upper part of the combs from the top of the glafs. The glafs of two parts is intended for the conveniency of taking the upper part off when full, and to be fucceeded by placing another. All glaffes are difficult to crawl up by the bees, occafioning extraordinary labour. In fmall glaffes efpecially, the crowds entering with their load, after much ftruggling find it not wanted there, nor perhaps in feveral others; and after all this toil are ob-liged to defcend with it into the hive. For this reafon, I have advifed none under a quart. But to thofe who are *not* anxious about quantity, fmall glaffes to their own tafte will be more pleafing. The greater the number of glaffes, or their magnitude, the greater fhould be the proportion of bees to fill them; or the box will contain moftly brood, and very little honey. An *addition* of a good fwarm or two is, in that cafe, neceffary.

Thofe who have large globular glaffes may

have

have them cut in two (by the glafs-cutters), and have a divifion board adapted to the under half, as directed for a central glafs.

It may be thought that, by the ufe of glaffes, the queen might be often difcovered: but the reverfe is true; fhe very feldom vifits them, having no bufinefs there, brood hardly ever being found therein. *Once* I had fome in a large globular glafs, owing to want of room in the box below. The drones often afcend in the glaffes to repofe themfelves. Glaffes do not *prevent* fwarming, for I have had fwarms rife, even after they were half filled.

MANAGEMENT.

To place glaffes over a box, fet them as in pl. 2. fig. 1. properly upon the *adapter*; flide the divider under the cover of the box, and fet the adapter and glaffes on the divider; then holding it fteady with the left hand, withdraw the divider by the right. Then cover the whole with a dark-coloured cloth. It is proper to omit infpecting them for two or three days. Any chafms that may

K happen

happen by the glaffes not fitting clofe, or by not being wide enough for the openings, may be covered by flips of bohea tea-cheft lead.

Glaffes may be fet on ftraw hives, by hav-ing a *circular* adapter fet over, as before mentioned; only be careful that the glaffes do not ftand too near the edges, fo as to prevent the body of a ftraw hive from fur-rounding them, or the ftraw cover from be-ing laid over; and which may be removed at pleafure for infpection.

No glaffes ought to be fet over ftocks, until a duplet is about half full, left the after-feafon fhould prove unfavourable for ftoring the boxes. No glaffes fhould be fet over *weak* ftocks. About the tenth of July glaffes fhould be taken off; but if duplets are well furnifhed, they may be fafely ad-mitted fo long as the bees continue to place honey therein. In cafe the bees of a duplet lie out, take the glaffes off, and raife the ftock on a nadir.

In *bad feafons*, glaffes cannot be filled without too much impoverifhing the ftock; probably to their utter ruin.

4

In

In four or five days of bad weather, the bees will feaft on the honey of the glaffes; to prevent which, take them off. But they muft *not* be put on again, on a favourable change; for they will take the reft of the honey; although, when done, they will re-fill them. Therefore put on frefh glaffes, with *empty* combs.

The glaffes fhould be taken off as faft as filled, and replaced by empty ones, or the openings covered with tea-cheft lead.

Two flips of double *tin*, each about half an inch wider than the bottom of the largeft glafs, are neceffary to take the glaffes off by. Slide one under the glafs to be feparated, and the other under the firft; then withdraw the upper tin, with the glafs thereon, while the other is kept clofe and fteady in its place, till an empty one is fet on. The glafs taken off is to be conveyed to a darkened room; and turning it on its fide, towards the light, the bees will fly directly thereto, and foon quit the glafs. If they do not, tapping on the fides with the hand, or blowing with a pair of bellows, will make them foon relinquifh it. Small glaffes are

K 2
to

ducing honey; otherwife the ftock muft be *raifed* on a nadir. Place the hive taken, a confiderable diftance from the ftock; and if in two or three hours the bees remain quiet, there is a prefumption of its having a queen, or brood, and it muft be fet on again. But when all the three hives appear crowded with bees, fo as to want more room, fet the hive that was taken, with its door as near as can be to the ftock door, fo as not to obftruct it; laying a flip of wood as a bridge from one to the other: and place an empty triplet on the ftock. The hive being placed thus near to the ftock, with its floor touching, will be efteemed ftill as one family, and the brood reared as fuch; and in about three weeks may be taken away. The brood in that time will be matured, and the cells filled with honey.

But in the interim, if an unufual crowd or difturbance, or crumbs of wax, are feen at the door, it is a token that the ftock bees, or fome others, have begun to pillage. Obferving this, take it directly to a dark room, and cover it up for a few hours: if then the bees are quietly efcaping, let it remain till morning,

morning, and then fume it, whatever be the
ſtate of the bees.

But if, after the triplet has been taken, the
ſtock is in confuſion, it is a ſign that the
queen was therein (though this ſeldom happens), and it muſt be replaced. This CAU-
TION is particularly neceſſary to be obſerved,
in reſpect of all hives when taken; as ſome-
time a *young queen* may reſide in one hive,
and the old one in another; or the old
queen may be in it herſelf. This is often
the caſe with duplets which have farina and
brood; and that even though the upper
door had been timely ſhut. *Generally*, when
a nadir is half full of combs, and the door of
the upper hive has been kept ſhut, the queen
begins to lay her eggs in the nadir; therefore, in about three weeks after, the brood
in the ſuperior hive will have been hatched,
and the cells filled with honey, and proper
for taking. No DUPLET is to be ſeparated
in autumn, unleſs the hive left, in all appearance, is quite full; then that which
ſeems moſt likely *not* to have the *queen* may
be taken; but if this cannot be determined,
it is moſt eligible to let *both* ſtand. The bees

K 4   will

will not be the worfe for having more food than is neceffary (if kept warm in winter) ; but may perifh by having too little, which may happen in a protracted bad fpring.

Bees will *not quit* a hive that has *brood*, whether upper or under, without fuming or driving. The following day after a hive has been feparated, if farina has been carried in, it fhews all is well ; but if not, return the hive that fails to the ftock again.

. When it happens that a *feparated hive has a queen*, and is well ftored, it may be kept, if fuch an increafe is wanted ; provided the ftock left has alfo a queen. But if, unfortunately, the ftock queen has been killed in the operation, reftore the hive taken, to its family.

The BROOD COMBS of hives taken, fhould be handled with great tendernefs and circumfpection, that none may be damaged or crufhed. Rather cut into the honey cells than into the brood ; and let them be kept warm, until they are fet over a ftock. Place them in an empty hive reverfed, without its cover ; the combs to be difpofed fo as to touch each other as little as poffible,

by

by placing flips of wood, half an inch in thickness, between, to give fufficient fpace for the young to be excluded, and for the paffage of the bees to nourifh them. At night fet them over the ftock they came from, or fome other that needs recruiting.

Deprivation fhould always be done in the evening, as foon as the bees are retired to reft; that there may be fufficient light leifurely to perform the operation.

The GENERAL TIME OF DEPRIVATION, or TAKING UP OF STOCKS, varies in different *counties*, according to their different temperatures; but about the latter end of Auguft is the ufual feafon.

Bees kept in *fingle hives* ought to be taken when honey-gathering begins to ceafe. This may be known by a diminution of activity in the bees (if not from bad weather); for, when this happens, they begin to feed on the hive honey, beginning with the unfealed or exterior cells firft. Therefore, the *longer* they are permitted to ftand, the *lefs honey* there will be in the hive, when taken; and that in proportion to the number

ber

ber of bees it contains; which at that time consume a great deal in a little time, and consequently prove an absolute *loss*. This is meant of stocks taken the *common way to be destroyed*. What hive honey they have eaten can be of no profit, when the bees themselves are soon to be killed.

But this is not the case in the STORY METHOD, the bees of which are always saved; and therefore no *disadvantage* can arise from their standing. For if a duplet that has stood be taken after having eaten a good part of the honey, it has saved a like quantity of the stock's, which they would have consumed, had they staid on.

At the usual season of deprivation there is *generally* much brood, whose preservation is of much importance : for, coming INTO BIRTH so late in the season, they will survive through the *next summer*, till the honey harvest terminates. *This brood, thus preserved, is of more worth than twenty times the number of promiscuous bees, taken from a stock, and incorporated with another;* even if the stock should prosper, which is very doubtful, as experience verifies.

It

It is surprising, that the *salvation of the brood* has never been noticed; although every one, on taking combs out at this season, might have obferved brood therein, in their feveral ftages of maggots, or nymphs, and often of eggs. Regardlefs thereof, they are mafhed indifcriminately with the honey-combs; thus greatly injuring its quality by fuch ill-judged conduct.

In the *ftoried method*, inftead of the general deprivation of *duplets in Auguft*, I apprehend, for the reafons above affigned, it will be eligible to defer it to the latter END OF SEPTEMBER, or the beginning of October; or till the weather is too cool for the bees to work much out; by which time all or moft of the brood will have been matured, and have left their cells, without the rifk of deftroying any of them: befides the advantage of performing the operation with more eafe, fafety, and fatisfaction; as at that time, from having neither brood nor princefs, the bees will quit the duplet, when feparated, in a few hours, of themfelves, without fuming.

In wet and cold feafons, honey-gathering

is

is very fcanty; a circumftance which leaves numerous vacant cells for the rearing brood, and thereby renders *deprivation* much later than ufual. For the hives may feel heavy, but it will not be from honey, but moftly from farina and brood (efpecially if the ftock is of two years ftanding); which may lead the apiator into a fatal error, as thinking the ftock *rich*, though in fact it may be *very poor*, and die of famine in the fpring. Stocks left double are not liable to this cafualty.

TO JUDGE OF THE WEIGHT AND CONDITION of a ftock fit for ftanding, befides the direction given before in this chapter, lift the ftock a little up: if it feels of a due weight, that is, about twenty pounds exclufive of the hive, it may be fafely concluded as fit to keep.

It will be ufeful on feveral occafions to *number and weigh* the hives and floors, before the bees are put in. By this means, any evening, by ftopping the hive door, they may be readily weighed, without any difturbance to the bees.

If any of the ftocks remain *trebled* till Auguft, take away the moft empty; for it

is

is advifeable, that the ftocks, in general, be reduced to duplets at this period. Thofe that have but few combs, are obvioufly to be taken. In a cold evening or morning, an affiftant may lift the hive high enough up, to permit the apiator to look underneath, which he may do with little danger, or dif- turbance to the bees. The doors of all duplets that feem moft vacant fhould be fhut.

All ftocks in common hives, that are light, fhould be taken; and none kept, unlefs about twenty pounds weight. Weak ftocks feldom furvive the next fpring; but, if by chance they do, turn to little account, not adequate to the trouble and expence of feed- ing. *One ftrong ftock* will be more produc- tive than four weak ones. Neverthelefs, in extraordinary fituations and feafons, they may yield tolerably well.

In favourable feafons three hives have been taken off, each yielding twenty pounds of combs, though in a fituation that was but middling.

Two or three cafts joined together, have accumulated honey very rapidly; while

o   their

their feeble neighbours, having few collectors, loft that fhort but precious opportunity.

It is beft to SEPARATE *boxes* about ten in the morning, when the greateft number of bees are out; as it can be done with more eafe and fecurity than in ftraw hives.

In fmall apiaries, the divider had better be fhoved under a hive the night before, and then the bees will be fo little difturbed as hardly to refent it.

When bees are terrified by the operation of deprivation, or other violence, they become regardlefs of their queen, till the panic has fubfided. At the feafon of deprivation, the light ftocks had better be incorporated, three or four, at difcretion, in a hive, and proportionally furnifhed with honey.

CHAP.

CHAP. XVIII.

OF PASTURAGE, OR BEE-FLOWERS.

A PLENTIFUL affortment of bee-flowers is a confideration that requires attention, if we defign to favour an ample production of honey. The *nearer* the pafturage is to the apiary, the *more* journies the bees can make in a day, and confequently the fooner they will be able to fill their hives.

The PRODUCT from a *large* fupply, but at a *fmall diftance*, and in a *temperate fituation*, even with the *common management*, will be fuperior to that of the moft fkilful in a *bad one*. On the contrary, with bad manage-ment, and with fcanty pafturage, and indif-ferent fituation, a very *trifling profit* can be expected.

BRITAIN in general is but thinly ftocked with bees. Few farmers in comparifon efteem them worth their notice; it is from

the

the attention of COTTAGERS we derive
the chief fupply of honey and wax. It will
be readily admitted, that a large number of
ftocks kept within a fmall circuit and in a
bad fituation, will be prejudicial to *that*
circuit, as being more than can be fupported
in affluence; and will neceffarily impoverifh
each other. The ftate of any particular
fituation may be known by the general product for feveral years together, and not
from one or two years only; but more certainly from what a very good feafon will
produce, which may be accounted as a
ftandard.

But there are *many fituations* capable of
feeding a much larger number of ftocks
than are to be found on them. However,
if the generality of farmers and cottagers
individually would keep a few ftocks, nearly
all the honey and wax this country could
produce might be collected. This would not
only benefit individuals, but might alfo be
of real *national* utility.

In many counties, cottagers' wages are
too low to enable them ever to purchafe a
fwarm or ftock of bees, efpecially if they
have

have families. It is a *prudent* and com-
mendable method they have *here*, of giving
credit for a fwarm; to be compenfated for
by the *firft good one* that it yields the next
year, and about a quart of honey for intereft.
I hope this *practice* will become GENERAL,
among thefe induftrious and ufeful people.
I flatter myfelf that the well-known bene-
volence of the BRITISH GENTRY will induce them to affign fome part of their influence to promote it.

Large HEATHS AND COMMONS, fur-
rounded with WOODS, are noted for being
abundantly productive: the *firft* abounding with wild thyme, and various other
flowers untouched by the fcythe; and the
other with a profufion of farina and honeydews. Heath and broom are very ferviceable, as continuing long and late in
bloom.

It is remarkable that the domeftic bees
are very *nice* in their feleCtions, and do not
rove from one fort of flowers to thofe of
another, indifcriminately. They are limited to a few kinds. Thofe of the moft
gaudy colours, and which afford the moft

L   refplendent

resplendent show, and agreeable odours, are mostly neglected by them, as hyacinths, jasmines, roses, honey-suckles, &c. while very small flowers, or those of little note, are to them plentiful sources of nectareous sweets.

A List of Bee Flowers.

Winter aconite, lauruftinus, hazel, snowdrops, crocus*, fallows, ofiers**, primrofes, hepaticas, violets, standard almonds, single wall-flowers*, onion, goofeberry, apricot and other fruit trees, laurel, turnips*, all the fpecies of braffica, or cabbage*, dwarf-almonds, rofemary*, ftrawberry, tulip, white-thorn, heath, gorfe, ftar of Bethlehem, borage*, viper's buglofs*, rafberry*, laburnum, tacamahacea*, columbine, barberry, bean, yellow lupine, fyringa, fweet-brier, muftard, tares, white clover**, cucumbers, greek valerian, fenna, French willows, holly-hock, ferpyllia or creeping lemon thyme**, capers, white poppies*, mignonette**, blackberries, lime-tree*, chefnut, mallows, hyffop, teazle, buck-

Q   wheat,

wheat, nasturtium, yellow vetches, saint-foin, alders, scabious, sun-flower, broom, Michaelmas daisies, winter savory, Jacob's beard, purple house-leek, tree-ivy; and a few others of less note.

Those marked with * are such as produce the greatest quantity of honey, or farina; and those with **, such as yield the *finest* honey. Some of them afford both honey and farina. They are ranked nearly in the order they blow.

Bees are most fond of spots where large quantities of their favourite flowers are to be found together. Fields of buck-wheat, or WHITE CLOVER, will be thronged with bees buzzing their joys, so as to be heard at a great distance; while plants that afford finer honey, but scattered here and there, will be neglected. When several forts of honey *flowers* grow near each other, they will only collect at *first* from those that furnish the best honey. For instance, if several species of thyme are planted together, they will prefer the creeping lemon thyme *only*, as long as its flowers last. In seasons of *scarcity*, they are obliged to take up with

L 2 species

species of a very inferior quality, and such
as they would despise at another time.

Besides the acquisition of honey, FARINA
is of great importance to make bees flourish.
It is the *dust* or flour found on the *stamina*
of flowers, and which contains much essen-
tial oil, visible to the naked eye on holly-
hocks. This precious concrete substance
the bees collect in little balls, on their hind
legs, or by the hair of their bodies. The
balls, on their return home, are struck off
from their legs, in its *crude* state or by
biting it off piecemeal, and are deposited in
their cells; other bees often assisting. Probably the farina of different colours may be
also as different in quality.

Its USE is *partly to feed themselves*, and
partly to nourish the young. *That* gathered
in *summer* is immediately swallowed, and
by their digestive faculties converted either into *food*, or *wax* wherewith to form
the combs, and which is discharged at their
mouths in a *soft* state; so well adapted is it
to its intended purpose. Therefore, when a
swarm is *newly* hived, little or no farina is
seen to be carried in; and a proof of this
may

may be *had*, by an attentive obfervation to boxes.

Wax is alfo drawn by the bees from the refinous and balfamic juices of trees; the purer fort from the leaves, and imported under the folds of their bellies. The bees that have the good fortune to acquire this precious article, on their arrival in the hive fhake themfelves very much, as though they had a difficulty to difengage it; and yet are impatient that others fhould do it for them. Tar and paint they will likewife load themfelves with, much to their prejudice. TACAMAHACCA yields refin fo abundantly, that the bees are very profufe of it, daubing the box windows fo much therewith as fcarcely to be feen through. Combs made with it are coarfe and clumfy. On the contrary, thofe made from white clover, or white poppies, are white and elegant.

WAX FROM AFRICA is chiefly drawn from refins of the nature of turpentine fubftances, and for that reafon bears a lefs price than Britifh, which is chiefly from flowers.

Sallows furnifh a larger quantity of fa-

rina

rina than moft other plants, and that as *early* as the bees have occafion for it. Rofemary is the firft aromatic plant that blows; it grows wild in fome parts of France, and is the caufe of that fuperiority for which the Narbonne honey is efteemed. Mignonette yields good honey, and is valuable for its long continuance in bloom, even till November. Beds of it near an apiary will be of advantage, as will edgings of creeping lemon thyme along the borders of the garden. Single wall flowers in plenty will be ferviceable. LIME TREES are not to be neglected about apiaries, ferving in a *double* capacity by their flowers, and by their *leaves* which are frequently covered with honeydews.

Neither *beans* nor ORCHARD TREES afford any great quantity of honey; as may be obferved by the ftocks in Herefordfhire, which, *though* abounding in orchards, is not more productive in honey than other counties. In contraft to this, the borders of Cambridgefhire and Hertfordfhire, and part of Hampfhire, abounding with large heaths, commons, and woods, are much more productive

ductive than any other part of the kingdom.
Farmers there have been known to keep
from a hundred to a hundred and fifty
stocks of bees.

Viper's buglofs is a plant much like borage. It is a very troublefome weed in
corn, among which it is found in many
places in great plenty ; and is fure to make
rich hives ; it has a biennial root, delights
in chalky or dry foils, and will grow on
old walls.

BUT BORAGE IS THE KING of bee-
flowers; it is annual, and blows all the
fummer, till the froft cuts it off. It affords
honey, even in cold and fhowery weather,
when other flowers *do not*, owing to the
flowers being pendulous. The feeds drop,
and fow themfelves ; the honey from it is
fine.

To find the quality of the honey from
any particular fpecies of flowers, if they are
in confiderable quantity, fet fmall glaffes
over a ftock at the time of their flowering,
and they will chiefly be filled with honey
of the predominant flavour.

Lavender and balm, though fine aro-
L 4 matics,

matics, yield little or no honey in our climate ; though they do in warmer countries.
In OURS, where wet and cold so often occur, the changes are so sudden (but generally not in all counties alike at the same time)
as to affect the flowers in the difference of
their products, suiting one sort, and not another. Lavender is a particular instance,
which is very abundant, and yields a large
quantity of honey late, when most others
have done.

VERY DRY SUMMERS are as unfavourable, in causing the flowers to fade and die
too speedily to yield much honey. Furze
or gorse, in many parts of Britain, the bees
collect from ; yet, in the vicinity of Pembroke, I have observed it to be entirely neglected by them : whereas, the quantity
here is so large in the hedges and fields,
that the product of honey would be very
great. Rape is very beneficial to bees, as
also turnip, and, as it is later in bloom, will
be serviceable when the other is gone.

Some flowers, it is probable, contain at
once all the honey they can furnish, and,
when deprived of that, yield no more,
though

though continuing in bloom much longer. As for inftance, white clover. I have feen fields of it covered with bees; but in two or three days they had not a fingle bee on them, although continuing in bloom, and the weather equally favourable.

In very fcanty feafons of honey-gathering, bees have been obferved to feed on mellow goofeberries, and ripe faccharine pears; but I believe none was carried in for ftore.

Bees do not fly to fo great a diftance as has been imagined for pafturage. The hotter the weather, and greater the profufion of flowers to be found on one fpot, the farther they will be allured to fly, and pafture thereon; perhaps a *mile*, or *a mile and a half*; but generally, it is moft probable, they do not exceed *half a mile*. When it is cool and windy, though they are fhort of provifions, they will perifh rather than fly beyond that diftance. Inftances of this I have feen in flocks in that condition, fituated in a large garden; which, on being removed to the fide of a large common, not

a mile

a mile diftant, prefently refumed their labours with vigour, and profpered,

WHERE LAND IS VERY CHEAP, it feems
reafonable to fuppofe, that it might be *culti-
vated* with fome of the moft productive of
bee-flowers ; fuch as white clover to ftand
and feed, rape, muftard, borage, viper's buglofs, ftrawberries, rafberries, or buck-wheat;
marfhy wet foils, with fallows, ofiers, or
lime-trees, which would be likely to prove
of more confiderable *advantage* for eftablifh-
ing a productive apiary, than to let fuch
lands remain covered only with four grafs,
rufhes, furze, and briers, and fuch like *un-
profitable* vegetables. Perhaps many perfons
will find their account in removing their
ftocks of bees to fields of clover, buck-
wheat, turnips, muftard, or heath, according as the flowers are earlier or later than
thofe of their own fituation,

CHAP.

CHAP. XIX.

OF HONEY DEWS.

HONEY dew has in general been erroneously fuppofed to be a dew that falls *indifcriminately* on all plants alike; whereas the true honey dew is an EXUDATION from the leaves of a few fpecies *only*, and that at a time when other dews do not exift. The trees and plants on which it is found, are the oak, maple, fycamore, lime, hazel, and blackberry; and fometimes, though very feldom, on cherry trees and currant bufhes.

Its *time* of appearance is about ten or eleven o'clock in the morning, and its duration about four or five hours, according as the fultry heat which produces it continues. Sometimes it is found as early as feven o'clock, and though the fun does not fhine out, if the preceding day and night have been fultry; or when the fun's rays

are

are reflected from clouds. It is not always found in the several species at one time, perhaps only on *one* in particular.

This substance is as transparent and as sweet as honey; in fact, it is honey. At times it resembles little globules; but more often appears on the leaves like a syrup, and mostly in the old ones.

The SEASON of its usual appearance is from the middle of June to the middle of July; but varies in different counties, and according as the weather is more or less favourable. In some years there is none at all. In general, when fruit is backward, so are honey dews; even so late as harvest. There have been instances of honey dews *two* months later than the usual time, owing to the wetness of the summer, and then but small in quantity. The stocks, when taken, were light, and those left mostly died of famine in the winter; except in the HEATH COUNTRIES, which blowing late, furnished honey that was but very ordinary, and barely adequate to their winter's wants.

When a honey dew is produced, the activity of the bees is violent and unremitting:

they

they almoft defert the hive to import it;
knowing its time of continuance to be of
fhort duration, and that on the weather
fuddenly changing it is entirely over.

While the trees are charged with it, the
bees are *as though* fwarming therein, buzzing
their joys in loud acclamations. But wo
and fmart to thofe who obftruct their fwift
defcent to their hives!

More honey will be collected in one
week from dews, than in *many* from flowers.
It is obvious, therefore, how *great muft be
the advantage of thofe kind of trees* in the vicinity, and from the *ftory method,* by which
the bees may (with care) never be at a lofs
for enlargement to beftow the treafure in.

CHAP.

CHAP. XX.

DISEASES OF BEES.

COLD, foggy, damp weather, in the winter, is very often fatal to bees: for *then* having no exercise they become subject to a purging, by which they are soon reduced very weak; and clustering together in a body soil each other, and thus contaminate the whole. The signs of this disease are small crumbs of wax about the door, or on the floor, with many *dead bees*, and much filth caked together, and, if of some time standing, mouldy, often concealing destructive wax-moths, &c. If the bees do not fly out, and appear as active as other stocks, it is a symptom that they are either *dead or starving*. In Hertfordshire I had many stocks affected; but in Pembrokeshire I never had *one* diseased; owing, I suppose,

to

to the ftrong and frequent ventilation of
SEA AIR, to which my fituation is expofed,
keeping the atmofphere always pure.

The difeafed ftocks are to be taken, as
foon as difcovered, into a warm room.
Brufh away the foulnefs from the edges of
the combs, cutting out the parts that are
mouldy or black: fet the hive at a moderate
diftance from the fire, which will revive the
bees that are feeble, or torpid: as foon as
they begin to move, pafs among them a
few drops of honied ale; tie a flight cloth
over the hive, that none may crawl out, and
let it remain three or four hours, to purify
the damp and foul exhalations. When the
bees are pretty well recovered, give them a
trough of honied ale in which the leaves of
rofemary have been infufed, and fet the hive
on a clean floor. Contract the door, fo as to
admit a little of the warm air. Let them
remain till next day. If then the bees are
few, or are ftill weakly, cover a dry floor
with afhes, place on that a little hay, or
ftraw, and fet the hive therein, conveying
it to its ufual ftand. Cover it well with
ftraw, bags, &c. and notice occafionally
whether

whether their condition may require further feeding; which fhould be given daily, if the hive is not fufficiently ftored with honey and farina.

When bees fall motionlefs to the bottom of the hive, it indicates that they are chilled with cold, or in a ftarving condition. To prevent a further deftruction, treat them as above, or fet them to a plentiful ftock.

Bees often fly in a defultory manner about the hives, bee-houfes, or dwelling houfe, in the *fpring*, with lamenting tones, as though wanting fomething: that *fomething is food*; for they are almoft famifhed. By obferving which of the ftocks has an unufual crowd at their door, the diftreffed hive may be difcovered. A frefh, dry and warm floor muft be given them; and they muft be immediately fed: the delay of a day may be a day too late.

When ftocks appear to be LIGHT, a daily feeding is indifpenfable, till a certainty of honey-gathering has commenced. Or a hive or box may be cut down to five inches, and filled with combs of honey, properly placed, which may laft them a long while.

The

The MORE BEES a hive contains, the greater their warmth, which caufes them the fooner to become active in the fpring; and accelerates the breeding of the queen, and the production of young. But the quicker alfo will the honey be exhaufted.

And this is the reafon why fo many *flocks perifh in the fpring*, when leaft thought of; *if* they were *fcantily ftored.* This confideration fhould operate as a ftrong inducement to keep NONE but rich ftocks.

A DEGREE OF COLD that fhall throw the few bees of a weak ftock into a ufeful lethargy, will not have that effect on one that is *populous.* On this principle the weak ftock will furvive; while the populous one perifhes, by confuming all the honey by the increafe of numbers; admitting both to have an *equality* of honey.

When *bees in cold weather* difengage themfelves from the body or clufter that is in the hives, or fly out, they are prefently chilled to death.

Thefe infects fuffer more through the inftability of our climate, in its frequent and

M   fudden

sudden transitions, than from a long continuance of frost. The milder the *winter and
spring* have been, the sooner their store is ex-
hausted; and if it was rather short at first,
the sooner the stock *dies;* or perchance it
may survive till the latter end of May.

The frequent FAILURE OF STOCKS has
in most counties been attributed to WITCH-
CRAFT, or other *superstitious* notions, instead of attributing them to their *true cause;*
badness of weather, or their owner's neglect,
or want of *skill.*

These causes operate alike in every article of husbandry; often blasting the fondest
expectations of the farmer. But he will not
be so absurd as to suppose that evil spirits, or
witchcraft, have any power to sport with
mortals, or their property, at pleasure; much
less that bees in particular should be victims
to their malice, more than sheep or cattle.
No! he patiently submits to the *Omnipotent
Disposer* of all events, from the destruction
of the ant-hill to the dissolution of mighty
empires.

To secure them from diseases, it will be
necessary

4

neceſſary (contrary to the common opinion)
to keep the hives *warm* in winter, by filling
the vacancies around and at top of the hives
with ſtraw; eſpecially box-hives. In ſnowy
weather, or very hard froſt, the door-ways
ſhould be wholly cloſed, which in ſuch a
ſeaſon will not be prejudicial; *provided* care
is taken to unſtop them immediately on the
weather changing; for as ſoon as that happens they will be very anxious to iſſue out
for freſh air, as alſo to empty themſelves.
Bees ſhould always be ſuffered to make their
exit, except as above, as they well know
what weather they can bear, and how long
to ſtay in it. It is beſt *not* to houſe bees in
winter; for when a mild day comes, they
will rejoice to take the air, which contributes
much to preſerve them in health.

The bees in winter ſhould be diſturbed as
little as poſſible.

When bees are long confined by ſevere
froſt, or rainy weather though in ſummer,
they grow diſeaſed for want of exerciſe, and
for want of emptying themſelves.

The regulation of the doors of the hives

M 2   ſhould

ſhould be proportionate to the weather and the populouſneſs.

The warmer the hives are kept the better. In cold ſprings the doors ſhould be ſhut at night, and opened in the morning; but be ſure that the bees have no exit, but of the hive, or it may prove their death.

CHAP.

CHAP. XXI.

OF FEEDING.

V ARIOUS have been the methods and materials for feeding bees in winter. I have found none more fuccefsful, cheap, or convenient, than SOFT BROWN SUGAR, that is not *grainy* ; a pound to half a pint of mild ale, diffolved over the fire. But as fugar is at the prefent very dear, honey may at this time fupply its place, though *inferior* for the purpofe. This compofition, which fhould be regulated to the confiftence of *fyrup*, comforts and ftrengthens the bees, preventing diforders, increafing their activity, and forwarding the brood, if given *plentifully in the fpring*.

It is to be adminiftered by means of TROUGHS made of joints of *elder*, angelica, or other kexes, flit down the middle, the pith and bark taken away, and reduced to

M 3   fuch

fuch a depth as eafily to pafs the door-ways
of the hives. Their length to be eight
inches, or fix at the *leaft*, and flatted a little
on the under fide, and the end clofed with
putty, or other cement. Thefe troughs, by
paffing *far* into the hive, enable the bees to
come down to feed, without danger from
the cold, which they would fuffer in coming
to feed at the door. They are alfo too narrow to fmother themfelves therein. The
larger the number of bees, fo much the larger
muft be their fupplies.

When STOCKS SHEW SIGNS of POVER-
TY, pufh into the hive a trough of the honeyed ale (by this term I always mean either honeyed or fugared ale, as may happen
to be cheapeft) in the evening; and if the
combs obftruct its entrance, pafs a long thin
knife to cut a free paffage. The next *evening* take another trough full, and, pulling
the empty one out, pufh in the full one;
and thus proceed as long as there is occafion. If ftocks do not come down to feed,
they fhould be taken into the houfe, and
fed.

Such a trough holds about half an ounce;
one

one of them is enough for any ſtock for a day and night. This I call PRIVATE FEED-ING. By this method they are prevented from feeding to excefs, which they are but too apt to do, when they have an abundant fupply at *once*; and thereby bring on a loofenefs, and prove both deſtructive and waſteful. Daily feeding, indeed, is more troublefome than giving a quantity at once; but the laſt is more expenſive, and not fo fafe. I fed, one winter, two very light ſtocks, through the dreary feafon of 1777, till the end of the enfuing May. By the means, and at the expence only of fixteen pounds of fugar, and one quart of ale, I faved my bees to flouriſh in profperity. Care ſhould be taken to place no feeding article on the *outſide*, or at the door-ways, as it will attract ſtrange bees, who may alfo become robbers, and ruin the ſtocks.

In fuch a difaſtrous feafon, a PUBLIC FEEDING may be *fubſtituted*, which is by taking an old empty comb (the deeper and harder the better), *filling* the cells on one fide with honeyed ale, and placing it on a hive-floor, and over that an empty hive, or pan;

and

and fetting it about the middle of the apiary.
The bees will foon flock about it in crowds,
and empty the comb: once in 24 hours
replenifh it. They will not come out to
feed in improper weather, though it continues for three or four days. Troughs of
food muft be fubftituted during bad weather. Nor muft *public feeding* be practifed
when other apiaries are pretty near, as the
bees of thofe will equally partake with the
owner's. The bees will entirely neglect
public feeding, as foon as *honey* can be obtained from flowers.

At a public feeding much quarrelling will
happen, between thofe who *are* feeding, and
others that cannot approach near enough to
partake for the great crowd; but it will be
unattended with mifchief—only mere box-
ing bouts, without ufing their *ftings* as in
fatal duels.

Feeding fhould not be *attempted*, until the
robbing feafon is over. If any ftocks before
that time are in diftrefs, they fhould have a
trough given them at night, and withdrawn
in the morning.

The *weighing*, or *poifing* of hives, in FE-
BRUARY,

BRUARY, to judge whether they require feeding, ought not to be deferred till after they have for fome time begun to breed; left the additional weight of them be miſtaken for that of honey, when perhaps there may not be a fpoonful in the hive, and the continual increafe of mouths produce the fpeedier famine.

Now and then a trough of food given to the ftocks as foon as farina is collected, will forward the queen's breeding, and likewife add much to invigorate the bees to greater activity in their labour.

I weighed a ftock November the 2d; it was then 29 lb. 3 oz. On February 26th, the weight was 24 lb. 1 oz,—Difference 5 lb. 2 oz. From November 2d to February 26th is 115 days (the weather mild), in which were confumed 5 lb. 2 oz. or 82 oz. which is but three quarters of an ounce per day.

On the 8th of December a ftock weighed 21 lb. 11 oz.; the 11th, 20 lb. 15 oz.; the 21ft, 20 lb. 8 oz. The difference from the 8th to the 11th, is 12 oz. i. e. almoft an ounce per day. From the 11th to the 31ft,
5 oz,

5 oz. is but half an ounce per day. The weather frosty the whole time.

In the firſt *thirteen* days the confumption was 12 oz. in the *ten* laſt only 5 oz. On further trials, I found the refults nearly fimilar.

From the whole I have been induced to conclude, that a trough holding about *half an ounce* of honeyed ale, daily adminiſtered, is a *fufficient fupport to any ſtock* while feeding is required.

Where the price of honey is higher than that of fugar, feeding will be of advantage, though the ſtocks do not need it. For what fugared ale they will confume, will be a proportional faving of fo much ſtock honey. Befides which, it will caufe thofe ſtocks, in the next feafon, to be the fooner fit for ſtorifying; and likewife, if it fhould be rigorous and long, the ſtocks, neverthelefs, would be rejoicing in plenty, while their neighbours would be ſtarving through fcarcity.

The feeding of bees, in fpring, is of great advantage to them, as it enlivens and ſtrengthens them, and ſtimulates their activity,

vity, caufing them to breed the earlier. A little good ale, with honey diffolved in it, will be very acceptable, even though they fhould be well provided.

Since the preceding fheets were written, I have found a very eligible method of feeding, by taking a half hive, or box, cutting combs of honey down to the proper depth, and placing them therein, on bars fimilar to thofe of the flock which they are to be fet over. Loofen the cover, thruft a divider under it, take it off, and then carefully fet the half box of combs upon the divider, and immediately withdraw it, and place a cover over the flock. The quantity of combs put in muft be proportionate to the wants of the bees, to the time of its application, and the nature of the feafon.

CHAP.

CHAP. XXII.

OF THEFTS AND WARS OF BEES.

THE bees of apiaries are often enemies
to each other, and wage deſtructive war,
compelled thereto by neceſſity.

The ROBBING SEASON is ſooner or later,
as the ſummer has been more or leſs favourable; but in general it happens in
March and *Auguſt.* That of March is but
ſeldom and trifling: in Auguſt very frequent and formidable. I once had a ſtock
attacked in this month, and again in October.

When ſwarms have been late, but numerous, or a bad ſeaſon has followed, it will
be a very dangerous time, and make it neceſſary to contract all the door-ways, as a
caution of ſecurity. A few bees will defend
a narrow paſs againſt a multitude.

As very bad ſeaſons often occur, which
prevent ſtocks from procuring ſufficient

honey

honey for their winter ftore; reduced to the choice of *ftarving* or *plundering*, thofe that are ftrong chiefly prefer the latter.

This being determined on, they fend spies to difcover the ftate of neighbouring ftocks; and fuch as are found to have but few bees, but much honey, are concluded to be proper objects for an attack.

A few of the fpies for feveral days dodge about the doors, trying to get in to obtain more certain knowledge of their ftrength and riches; but are driven away by the powerful ftocks, who then plant guards at their door, which the weak ftocks do not, and therefore are the firft to be affaulted. The next day they return in force, and begin a violent fiege; and a defperate conflict enfues, both within and without the hive, neither fide giving quarter.

The ftouteft warriors make a defperate attempt, and rufh forward and feize the queen; knowing that, by difpatching her, inftant victory is the confequence; for the affaulted bees always defift, and join the victors, the moment they are apprifed of their *queen's death,* become as one fraternity,

and

and assist to carry their *own* treasure to their new habitation. But in case the queen is protected, they fight on with rage and fury, and death and pillage soon destroy the stock.

As soon as strange bees are perceived, contract the doors to half an inch; and when an attack is actually begun, stop the doors of *all* the stocks; taking care that no admission can be had, at any chasms, into the hives, till a little before dark; and then open all the doors, and the thieves will rush out and fly home, and the true bees, that were excluded, will enter in.

About an hour after lift the stock up: if it is *not heavy*, it must be taken and set over another stock, by *fuming*. But if heavy, and not much plundered, take it to a dark out-house, and keep it there two or three days confined, with some admission of air.

Very early in the morning shut all the doors, and post a person near the stocks that were most likely to be assaulted, with a kind of battledore, of flight wood, in his or her hand, with which to strike all the bees down that shall appear, and tread upon them. Continue this *sport* as long as any approach,

and

and in a few hours these formidable despe-
radoes will be destroyed. It will be finished
about noon. As the apiator's bees *are all*
confined, those killed are sure to be robbers
only; but if they should happen not *all* to
be killed in one day, keep them still confined,
till night, and finish the work next day.

When stocks do not shew resentment
against the attempts of the spies, and there-
upon keep guard, it is a very suspicious sign
of their weakness or poverty. They should
be roused to anger by thrusting some twigs
into the door-way, which will urge them to
revenge, and to guard their door.

But if *not*, take the hive, or the enemy will
be sure to strip it. The guard at the doors
will continue two or three weeks, if robbers
are about.

But when robbers find all the stocks upon
their guard, and courageous, after essays for
two or three days, they will desist, and retreat
to other apiaries in the neighbourhood more
favourable to their design.

When a stock has been assaulted, and all
on a sudden becomes quiet, with great crowds
of bees passing to and fro, it denotes the
death

death of the queen; on which immediately
clofe the door, and take the hive into a dark
room; and in the evening unftop the door,
when the ftrange bees will take wing for
their own home. Then take the combs
out, and *fave the brood*; or if the honey or
brood be fmall in quantity, *referve* the
hive as, it was left, to *fuper-hive* a ftock next
year, or to put a fwarm in.

As soon as strange bees are feen
about the ftocks, it will be prudent, *if* there
are any weak ftocks, to unite feveral into a
well-ftored hive of honey, which will not
only roufe the courage of the bees, but render them too powerful to be conquered.

The bees of good ftocks are always very
irritable and *revengeful,* whenever invaders
are on the fcout; nor will they let their fa-
miliar friend the apiator at that time approach them.

CHAP.

CHAP. XXIII.

ENEMIES OF BEES.

MANY, various, and powerful are the enemies and deftroyers of thefe induftrious and beneficial infects. But a little timely care and attention would prevent or greatly diminifh their depredations.

The *Wood-pecker*, or Tree-creeper, feizes the bees as they are gathering farina off the fallows in the fpring. *Robins* and *fparrows* will boldly wait at the hive door, and catch them as they come out; and fundry birds *feize* them in their *flight*.

Poultry are very prejudicial to bees, by catching them as they pafs in or out of their hives; and their dung is a great nuifance to them.

Mice get into the hives by the large and deep gaps made for door-ways in common hives readily admitting them in winter, to the deftruction of the ftock. They often alfo make a lodgment and breed under the

N crown

crown of the hackel, and eat their way through the top of the hive, to the ruin of the flock. Inſpection ſhould be taken to prevent it; and traps ſet to catch the mice. A good cat, bred in the garden, would devour them. The doors of the hives ſhould be made too low for a mouſe to enter, but at leaſt three inches wide.

The *wax-moth* is but little noticed, or even ſuſpected of being, as it is, a very dangerous enemy, deſtroying many ſtocks in a concealed manner. The mother moth lays her eggs about the ſkirts of the hive, if ſhe cannot *elude* the vigilance of the bees, to lay them in the *inſide*. She ſpins a cloſe and ſtrong web to defend the young, who burrow in the floors, and progreſſively conſume the combs, to the total deſtruction of the bees.

OLD STRAW HIVES, or DECAYED FLOORS, are very favourable to their depredations. Frequent ſhifting the hives, and cleaning the floors, will prevent the evil; and will guard againſt other diminutive enemies, as ear-wigs, wood-lice, and ants. The neſts of theſe ſhould be deſtroyed; or platters of

honey

honey and water, covered with brown paper, with many holes, which the ants may pass, but not the bees, and tied close round, will entice them to their destruction. *Spiders'* webs should not be suffered about an apiary.

LARGE SLUGS, or snails without shells, creep into the hives in wet weather; and are troublesome to the bees, by hindering their labour, and soiling the hive by their excrements; causing the bees to be very fractious; but they neither consume the honey nor wax; and generally, sooner or later, blunder their way out again: for I very seldom found one in taking a hive up, though I have often seen four or five at a time in boxes. By chance, they sometimes lie against the door-way, and stop it quite up; which may be soon discovered, by the bees not being able to enter. They may be taken out by a sharp-pointed wire in the form of a hook.

WASPS are much more destructive to stocks than their other adversaries, by their superior strength and prodigious numbers; especially in a year favourable to their breeding. They are most numerous in July and

August.   Soon after that the workers die;
but the mothers survive the winter, and
commence breeding about April.   But if
cold and wet weather enfues,.greater part of
the brood are ftarved ;· becaufe the workers
cannot fly out for forage, and wafps never
lay up any ftore.   Wet is very injurious to
their nefts ; and therefore, in a long feafon of
heavy rain, few wafps will appear till September.   But a mild winter, fucceeded by
a hot fpring, will fo favour the increafe of
wafps, that, without the greateft vigilance,
many ftocks will fall victims to their power.

One wafp is a match for three bees.   They
are very bold, and frequently encounter the
moft evident danger, undauntedly oppofing
a hoft of bees, to filch a belly-full of honey.
Therefore, when cold weather fets in, knowing that the bees keep no guard then,
great numbers get quietly in, and carry off
abundance of honey; and having once
tafted of the fweets, they will not defift
till they poffefs the whole.   *Perhaps* the
fame method of deftroying them, in this
cafe, as directed for bee robbers, would
prove as effectual againft wafps.

When

When wafps are feen dodging about the
hives, contract the doors to half an inch;
and fhould the bees be negligent in guard-
ing their doors, roufe them to anger by agi-
tating twigs within the door of the hive,
which will induce them to guard, and affail
the wafps.

In the fpring the *mother wafps* may be
feen about old timber, with the fplints of
which they compofe their nefts. On the
bloffoms of goofeberries and rafberries they
will be found often, and may eafily be,
knocked down and deftroyed. Their death,
at *that time*, will prevent a like number of
nefts from exifting the next fummer. A
neft of wafps, naturalifts inform us, confifts
of thirty thoufand.

Their *nefts* fhould be fought for by children; who, for a trifle, would feek, and give
information of them. Effectually to deftroy
a neft: In the evening, when the wafps have
done labour, repair to the place, and ftop
all the holes of their egrefs or regrefs. Introduce a squib into the chief paffage, and,
inftantly ftopping it with a fod, &c. they
will prefently be fuffocated. Dig the neft

up,

up, and burn it. Perhaps a wild-fire, of damp gun-powder, placed on a piece of wood, and introduced, would anfwer the fame purpofe.

Another way is, to make a hole in the top of their neft (ftopping all the others), and then pouring a quantity of boiling water down. This plan might be fubftituted for any method by means of fire, where gun-powder might be dangerous.

I have known wafps fo abundant, that in one feafon they deftroyed ten ftocks, in one apiary, out of twelve. A few fhillings, prudently diftributed, probably would have prevented this difafter, and diminifhed their nefts next year.

HORNETS, in the fpring, will watch the bees as they iffue from the hives. When they are feen about the hives, they fhould be knocked down and trodden upon. They may be trepanned, by placing an empty hive, with its infide fmeared with honey, among the ftocks. Allured by this, the mother hornets will begin to build therein. In the evening lift up the hive, which may be done with fafety, if the mother is there ;
then

then set it down again, and in about half an
hour after, have a veffel with water ready;
take the hive and plunge it a little way into
the water; then ftrike fmartly on the top of
the hive, and the hornets will fall into the
water, and by a pair of tongs may be crufhed
to death. OR, the hive may be clofely
ftopped up till morning; and then taking it
into a room, raife the edge next the window:
the hornet will fly directly thereto, and may
readily be cut in two by fciffars, crufhed, or
knocked down.

Their nefts are ufually hung on the raft-
ers, beams, or roofs of barns, or out-houfes,
or fixed in hollow trees. They refemble a
globe of brownifh paper.

The NEST MAY BE TAKEN by prepar-
ing a large-mouthed bag, with a running
ftring, to draw the mouth clofe. On a rainy
day, or in an evening, put on the bee-drefs,
and with great ftilnefs approach the neft,
and draw the bag gently over it, inftantly
pulling the mouth fo clofe, that not a hornet
may efcape. Separate it from the parts
it may be attached to, by a long knife,
plunge it into a proper depth of water, and

N 4 let

let it remain till morning. By this time the
hornets will be motionlefs; then taking the
bag out, tread upon it, to crufh the neft flat.
Turn the neft out upon a parcel of ftraw,
which being lighted, will of courfe burn
them; for the water will not kill them; and
they will revive. But if poultry are at hand,
the cakes of brood may be taken up by a pair
of tongs, and laid before the poultry, and
they will foon devour the young as a deli-
cious feaft. The fame may be done with
the brood from wafps' nefts.

CHAP.

CHAP. XXIV.

EXTRACTION OF HONEY AND WAX.

THE hives fhould be kept in a warm room, till the combs are taken out; fince the honey will drain out the fooner while in a fluid ftate. Turn the hive upfide down, cut through the *ends* of the fpleets clofe to the hive; then with a broad but thin knife cut through the *edges* of all the *combs*, clofe to the hive, and lift it on a clean board, or fhallow difh, having firft taken off the ftraw cover. Then, by a chiffel or wedge, force the *body* of the hive up, which will be effected if the ends of the combs have been properly loofened; and by this means the combs will all be preferved in their natural order, as fixed at their *tops* to the frame of bars : difengage them *fingly* with the knife, cutting a notch out of each, where it is faftened to the fpleet (which keeps the combs

all

all in their places) till the laſt is diſengaged.
The combs being thus preſerved *entire*, lay
them in a cleanly manner on diſhes, and
ſlicing off the cover of thoſe ſealed up, let
the honey run out.

The combs of *common hives* cannot be
taken out *whole* (though ſpleeted according
to my directions) without an iron *inſtrument*
in form of an L. The ſhaft to be that of the
depth of the hives, excluſive of the wooden
handle; the ſhort foot is to be two inches
long, and half an inch wide, made ſharp to cut
both ways; the handle, of wood, four-ſquare.
This is to be paſſed down between the
combs to the hive top; then turning the
inſtrument half round, and drawing it to
you, the combs will be diſengaged from
their faſtening to the top of the hive.

Proceed then to looſen them from the
ſides, &c. as above directed, and they may be
taken out without cruſhing and breaking
them to pieces.

The taking out the COMBS WHOLE, or
nearly ſo, is of great advantage to the pre-
ſervation of the *brood*, and the purity of the
honey; which may by theſe means be extracted

tracted without mixing the fluids of brood, or dead bees, or any other heterogeneous matter with it. '

Carefully feparate and preferve the parts of *empty* virgin combs by themfelves, for placing in glaffes; and thofe that are black, droffy, or charged with farina or *dead* brood, keep apart.

The FINE COMBS are to be drained and melted by themfelves, as being free from any alloy. They may be mafhed by the hands, and put upon hair fieves, as being pure virgin honey.

The parts of combs that have brood or farina in them, are to be cut out rather beyond their extent, to guard againft the chance of cutting among the brood cells. The *inferior* combs muft have all their defiled parts cut out, and *then* be fqueezed over fieves, or bolting cloths ftretched over fticks, laid over dripping or other wide pans, &c. and placed at a proper diftance from the fire, or in a room that has one, for the more fpeedy running of the honey. But for greater expedition, in large apiaries *preffes* are ufed. The pots of honey fhould not be

tied

tied down till a few days after their filling,
that the small particles of wax or other foreign matter may rise to the top, and be
taken off.

The portions of combs that were laid
aside as very impure, but containing honey,
may be cut, and thrown into water, to make
ordinary mead; or brewed with malt, to
make what is in Pembrokeshire called brag-
get; or else set before the bees on broad
dishes, &c. but spread thin to prevent the
bees from stifling themselves thereby; as
may likewise the refuse combs after draining,
and afterwards the vessels; first strewing
over them hay, grass, or herbs, to keep the
bees from being soiled. They will lick
up every drop of honey. It should be set
before them towards the evening. But if it
is not carefully done, many bees will suffer
by quarrelling; so that I think small mead
had better be made of them.

Having thus drained the honey from the
combs, BOIL THE FINE COMBS by themselves, with a sufficiency of water to keep
them floating, till they are thoroughly melted.

A *three-cornered* BAG of strong linen
cloth,

cloth, tapering to a point, is to be prepared,
which is to be held by an affiftant over a
tub of cold water, while the operator pours
the melted combs into the bag; *inftantly*
draw the top of the bag clofe by a ftring,
and let two perfons prefs it ftrongly downwards, between two ftrong fticks tied together at one end like a flail. Do this repeatedly down the fides of the bag till no more
wax iffues through. When the wax is cold,
it is to be taken from the water, and *re-
melted* with very little water, merely fufficient to prevent burning. As it boils, take
the fcum off as long as any rifes, and pour
it into proper veffels.

Thofe that are narrower at bottom than
top (the moft fo) are to be preferred. Rinfing the veffels and all the inftruments with
cold water *firft*, prevents the wax from ftick-
ing thereto.

The veffels or moulds for wax are to
be placed fo as to have the warmth of the
fire, with a cloth over them, that the *wax*
may cool *gradually*, or it will crack. When
quite cold, turn out the cakes of wax, and
pare off all the dregs that may appear on the

top,

top, or bottom, that it may be clear and marketable. The dregs that are pared off may be re-melted, and will yield a little more wax.

Inftead of perfons to hold the *bag*, which is fatiguing, it may be flung upon a ftrong ftaff, with the ends refting on the backs of two chairs, &c.

Or a four-legged *frame* might be more eligible; high and wide enough to admit a tub of water in the infide; and with ftrong pegs fixed on the top, at proper diftances, for fuftaining the bag in the middle of the frame. The bag is to have a running ftring to draw the mouth together.

The veffels in which wax is boiled ought to be confiderably larger than the matter contained; for when the wax boils, it very fuddenly rifes to a great height, and may prove of DANGEROUS CONSEQUENCE.

A more expeditious method of extracting the wax from FINE combs is, by boiling them alone. Prefs them flightly down, ufe very little water, keep them ftirring till the fcum rifes, which take off as long as any rifes; but when only froth appears, blow

. 3   that

that afide. When perfectly diffolved pour it into proper moulds, and fet it near the fire, covered over, till cold. On turning it out, the fmall quantity of impurities which has fubfided to the bottom, is to be pared off.

If the cake of wax fhould by chance feem difcoloured, re-boil it again without water.

Wax, when taken off the fire, cools nearly as foon as metals; therefore the procefs fhould be executed as expeditioufly as poffible, or a lefs quantity of wax will pafs through the ftrainers.

If combs are kept a confiderable time, without being melted, they will moulder and rot, or the wax-moth will breed among them, and devour the greateft part, and pefter the whole apiary.

A hive of three pecks, well filled with full honey combs, of two years ftanding, will yield in general 25 lb. of honey, and not more than 2 lb. of wax. The average run of common hives is 15 lb. of honey, and 1 lb. of wax.

CHAP.

CHAP. XXV.

CHARACTERISTIC OBSERVATIONS ON HONEY.

HONEY varies in quality, according to the nature of the flowers from which it is gathered.

That from aromatic plants is the beſt. But often, through very bad weather, the bees are neceſſitated to collect from flowers of very ordinary and diſagreeable qualities ; cauſing the honey of particular ſituations to be bad, while in other counties at the ſame time (the weather having been more favourable) the honey was of a very ſuperior degree of excellence.

VIRGIN COMBS *are ſuppoſed* to contain none but honey of the fineſt quality ; yet, if the above principle be true, ſuch may be ordinary. *All* combs taken from SWARMS are commonly eſteemed virgin: but this is an error, if by virgin is meant the pureſt and

I   beſt.

beft. For every comb, or part of a comb, that has had *farina* or *brood* in it, is thereby rendered impure, fo that fo much of any comb or combs, whether of *fwarms* or *ftocks*, thus circumftanced, is not virgin.

This is evident from the *fwarms* breeding through the fummer equally with the ftocks, and their combs being equally charged with brood and farina. The WAX from fuch combs will indeed be *finer*, and in *greater quantity* than that of ftocks. The continued ufe of the cells in breeding, firft foils them, and at length renders them impure ; but the cells where nothing but honey has been depofited, and which, when full, have been fealed over, are certainly MOST PURE. As to quality, *that* of virgin combs may be as *ordinary* as that of ftocks, if both were gathered at the fame time, and from the fame kind of flowers.

The *older combs* are the weightieft; for the bees will cement the *fkins* of the different breeds of maggots to the fides of the cells, to ftrengthen them, till at laft they become as ftiff as brown paper.

Old farina and other matters are continually

O

nually an increafing addition to their weight
and confiftence, fo as with difficulty to be
feparated from the *real* wax, even by a
long boiling, and then but partially. Honey
depofited in fuch old combs, neceffarily re-
ceives a tinge, tafte, and fome impurities
from them. However, *parts* of fome combs
in old hives, that have been *lately* made, may
be virgin.

It is for the reafons above, that a hive of
ftale combs, though bulky and weighty, dif-
appoints the expectation, by producing, in
general, only one pound of wax.

The honey generally brought to the *London* and other markets is moftly foul, and of a
coarfe quality, from the caufes above ftated,
as well as from the carelefs and uncleanly
manner by which it is *extracted.* The common method is, by taking the combs out of
the hives by piece-meal, *indifcriminately*, and
mafhing them, dead bees, brood, farina, and
drofs all together; which muft needs render
it an heterogeneous mafs, of a difagreeable
and often naufeous tafte, and unwholefome
in quality.

For

For fieves exclude only the groffer parts; but the fluids of the maggots and dead bees, with many other impure particles, remain intimately incorporated with the honey. By this unfkilful management a very valuable and falutary article of diet and medicine has been rendered difgufting and inelegant.

With fubmiffion, I would recommend to the nobility and gentry to purchafe none but COMBS of honey, to be drained at home: Sophiftications and impurities would then be avoided, and fuch combs might be felected as are fine, or according to their own fancy. Were this condition *infifted* upon, the markets would foon abound with COMBS of honey inftead of pots. *The introduction of fuch a cuftom* muft depend on the patronage of the gentry; without which fo ufeful an improvement will not be likely to take root.

Doubtlefs the price muft be regulated according to the quality of the combs, as in fugar and other articles.

Another benefit may arife from it, the promoting of the IMPROVED MANAGE-

MENT

MENT of bees; for as in the common
method few, *very few*, fine combs *can* be
produced, compared to *that* of ftorifying;
the peafantry would thereby by degrees be
influenced to adopt it.

The comparative tafte and fragrancy of
honey are the beft criterions to judge. of its
excellency.   In cold weather it grows hard
and grainy; fome forts are of a whitifh colour, as that gathered from white clover.   In
warm weather, or in warm rooms, it will ferment, and grow acid.   In fome years it is
naturally very glutinous and thick, to what
it is in others.

Honey, as partaking of acidulous and
faline parts, ought not to be kept in veffels
glazed with lead, as all coarfe ware is, but
in ftone: for though its effects may not be
felt by the ftrong, it may prove detrimental
to the weak and delicate.

Sometimes a white mealy matter will feparate, and concrete about pots of honey,
which is a real meal or farina that the bees
digeft with their honey.   The *white* at-
tracts the notice, from being the more con-
fpicuous.

Honey

Honey may be clarified by putting it into a bowl, and fetting that in water over a fire. When it boils, part of the impurities will rife to the top, and is to be fkimmed off. The heat, in this procefs, however, takes off from its fragrancy, and, if properly extracted, it is not neceffary.

It may be thought that honey retains the virtues of the flowers from which it is gathered. This may be true in a degree, efpecially of aromatics; but as it is gathered from a variety of different flowers, of various qualities, the honey muft partake of that of the aggregate. We find, whatever flowers it may have been collected from, it ftill retains its difagreeing quality (though otherwife diverfified by flavour and colour), and, if expofed much to fire, lofes its fine *fmell* and *tafte*. Neverthelefs, whether it could be deprived of its difagreeing quality, and made as agreeable as fugar, without a diminution of its medicinal virtues, merits the confideration of the chemift.

The *heating* and *griping* properties of honey probably arife from its *effential oil*, with which farina largely abounds; its detergent

and

and faponaceous qualities, from a fixed alkaline falt, combined with the effential oil.

It is wonderful, amidft the great chemical difcoveries of this age, that this beneficial article fhould never have been thought of importance enough to obtain an analyfis; by which a procefs might be deduced, to free it from its offending qualities, without impairing its medicinal ones. Probably *fuch a refinement as is ufed with fugar* might produce the effect; though with the lofs of its fragrancy and fine tafte.

Bees will not feed on candied honey, nor fyrup formed of rough-grained fugar, but fuck up the liquid part, and leave the granules behind.

CHAP.

CHAP. XXVI.

TO MAKE MEAD.

To every gallon of water add three pounds and a half of honey. Boil it as long as any scum arises, which skim off. If it boils longer, the fermentation will not succeed so well, nor will the liquor prove so fine.

Pour it into a cooler: at a proper degree of heat, put in a slice of bread toasted hard on both sides, covered with fresh yeast, and with it a little lemon peel, or any other pleasant-tasted substance. Set it in a warm place, and cover it from the cold air. When it has fermented two or three days, tun it up, and slightly cover the bung-hole; *taste* it every day, till it is found to have a *vinous* flavour and smell. Bung it then slightly; and when it appears to have entirely done fermenting, stop it quite down.

If

If another fermentation fhould be per-ceived, leave the vent peg out for fome days. Having ftood fix months, if it is fine, bottle it; if *not*, draw it off the lees, drain them out, without rinfing the cafk, and return the liquor into it. Then take a long two ounce phial (fuch as Bateman's drops or Godfrey's cordial are ufually put in), put therein a quarter part of chalk in fmall bits, and to it a quarter of water: then tying round the neck a piece of thread or twine, let it down into the cafk, till its top is on a level with the bung-hole; when pour in about a quarter part of the meafure of the phial of WEAK fpirit of vi-triol, and inftantly let it down far enough for the bung to go in; but not fo low as for any of the liquor to pafs into the phial. Hold the ftring till the bung is faft in, to fe-cure the phial from flipping down.

Care muft be taken, when the bung is to be taken out, to fecure the ftring that the phial may not fink into the liquor. The quantity ftated is enough for nine gallons.

The fixed air generated from the phial will gradually pafs into the liquor, and not

only

only fine, but tend greatly to preferve it from acidity, and give it the fparkling quality of champagne; taking off the difagreeable lufcioufnefs fo common in mead. Having ftood four or fix months longer, it will be fit to bottle. If any part of the fpirit fhould rife with the fixed air, or by other means get into the liquor, it will be harmlefs; being often prefcribed in medicine. Perhaps fharp vinegar may anfwer as well.

I conjecture malt and other fermented liquors will equally be benefited by a fimilar ufe of the vitriolic acid. Perhaps, if ufed at FIRST with the ferment, it would anfwer the purpofe much fooner.

Honey is preferable to *fugar* for making domeftic wines, giving the lightnefs, cordiality, and vinofity of foreign wines.

Mead may be flavoured by rafberries, ourrants, &c. by a proper quantity of fuch articles, that have been preferved with honey or fugar, being infufed into the liquors when fet to ferment. A fmall quantity will *then* flavour a much larger quantity, than a much larger if boiled in the liquor at firft.

If the liquor *ferments too long* after it is

tunned,

tunned, brimſtone thrown on a few live coals, and ſet under the caſk, will preſently reſtrain it, or any other fermenting liquor.

To *promote fermentation*, care muſt be had that the caſks be not ſhook, and that they be kept warm, excluding much air or light; and with the bung-hole but looſely covered.

For conducting the fermentation with ſucceſs, the rule is to ſtop it at the vinous ſtate, before it commences to be acidulous: for, if not fermented enough, it will be fóul, mawkiſh, and not keep; if too much, it will then turn ſour.

The practice of vintners is to ſcent their caſks with the match, viz. for a pipe take four ounces of brimſtone, of burned alum one ounce, put in a pipkin, and held over a chaffing diſh of coals till the brimſtone is melted and runs. Slips of canvas or coarſe linen are then dipped into it, and the powders of nutmegs, cloves, and corianders, inſtantly ſprinkled on them, and then fired, and let down at the bung-hole, and the fumes kept within the veſſel as much as poſſible.

This prevents the ropineſs of liquors, and a diſſipation of ſpirits, and conſequent weak-
.neſs,

neſs, ariſing from the imbibing quality of *new* caſks.

When vinous liquors become flat, they may be reſtored with ſpirit of wine, and with raiſins and ſugar, or honey. Theſe articles ſoon render them briſk, and ſparkling, and reſtore their ſtrength.

The juice of elder-berries will communicate a fine claret colour and taſte. An agreeable roughneſs may be alſo given by the juice of ripe ſloes.

CHAP.

CHAP. XXVII.

A SUMMARY OF MONTHLY MANAGEMENT.

As the moſt natural, it will be proper to begin our BEE YEAR with

Octtober.

This month requires no other ſuperintendance, than ſome caſual obſervations, viz. that the ſtocks are not attacked by robbers (for this, though not common in this month, ſometimes happens) ; and that no inſects or vermin harbour about the hives.

November.

It is proper to clean the floors, or rather to exchange them for clean and warm ones. Cover boxes, eſpecially about the tops, with matts or ſtraw. If any ſtocks are light, feed them, which in this caſe muſt be continued

through

through the other cold months. Clear away cobwebs, weeds, and vermin.

December

. Requires a continuation of the fame precautions. If an uncommonly fevere froft happens, fecure them effectually with coverings, and clofe the doorways ; leaving only a very fmall vacancy for frefh air. And in fnowy weather it is to be attended to that no bees may come out.

January.

The fame directions are to be obferved as for the two preceding months.

February.

- Feel the weight of the ftocks : thofe that feel light feed daily, till honey-gathering arrives. If two or three troughs of honied ale are given each of the ftocks in this and the following month, it will contribute to forward the brood.

March.

As foon as the bees begin to work brifkly, the floors fhould be again fhifted, and

every

every annoyance about the hive taken away. Early in the morning will be the propereſt time.

Thoſe ſtocks that appear to be very numerous (if the weather be *mild*) ſhould be duplicated.

April.

The flowers in this month are often replete with honey, and the ſtocks with young bees, ſo that ſwarms are ſometimes emitted; to which attention is to be given.

Through the windows of boxes may be ſeen whether honey is carried in, and then feeding may ceaſe, unleſs on a change to bad weather. Obſerve to *double* all the ſtocks.

May.

The weather in this month is moſtly very changeable, ſo that light ſtocks require ſtill to be fed, when it is unfavourable, even to the laſt day of its continuance. If the weather is hot, take off the additional coverings put on in the other months. Be ſure now to let the bees have a plenitude of room for breeding; better too much than too little. But if the weather is cold, mifty, and damp for feveral days, and not attended to, famine may be the confequence.

This month generally furnifhes many fwarms: therefore conftant watching is requifite from eight till three; or otherwife great part of the prime fwarms will efcape.

June.

By tapping on the fides of the duplets, it may be known whether they want the addition of a triplet. About the latter end of this month it is likely it may be neceffary to take off fome triplets, and to fet nadir hives under.

Be very circumfpect with regard to the ftocks that have *not* fwarmed.

July.

Swarms often rife till the end of this month; and therefore the bees muft be watched till all the hives have fent out their prime fwarms. Take hives off, and place nadirs under, as often as may be requifite.

About

About the tenth, the upper doorways of duplets muft be ftopped.

If the weather is fo hot as to endanger the melting of the combs, give the hives as much air as poffible, and fcreen them from the fun, and pour water upon the ground around them.

Auguft.

This is a dangerous month for robbing. Therefore an obfervation muft be had every day, to fee whether the hives are affaulted. By neglecting that, many ftocks are frequently loft. Wafps are to be guarded againft.

About the latter end of this month is the ufual feafon of *general deprivation*, or taking up of ftocks. Inftead of taking off duplets in this month, it would be better (I think) to defer it till the latter end of the next month, or beginning of October.

September.

No other attention is required than a cafual caft of the eye, to fee that the ftocks are not annoyed by robbers, or vermin.

PART

PART II.

INSTRUCTIONS

FOR

PERFORMING

THE

OPERATIONS.

P

GENERAL RULES.

I.

To put on the bee-drefs whenever an operation is to be performed; for although not always neceffary, yet it will be prudent to be prepared againft the worft, efpecially for the unexperienced. For a foot may flip, or an accident happen that no human forefight could be apprifed of. Great care fhould be taken after the drefs is off, of coming near the bees, as they will be eager to fting, for three or four days, though the perfon be at a confiderable diftance.

II.

Before any operation on a ftock, ftop or fhut the door-ways, and be fure to unftop them as foon as it is over, unlefs where it

P 2   is

is otherwife directed. The beft material, as well for this purpofe, as for ftopping crevices, is long fhaggy *mofs*, found on banks under hedges.

III.

Though the operations are directed to be performed in the morning early, or in the evening as foon as the bees are all at home; yet by the ufe of the *dividers* they may be done at any hour; in cloudy mizzling days; when the bees are out at their labour, or have been previoufly fhut in very early in the morning.

No. I.

No. I.

INSTRUMENTS OF FUMIGATION

ARE, firſt, a Box, pl. i. fig. 4. adapted to this purpoſe, of the exact *ſize* of the boxes in uſe. It muſt have a cloſe bottom, nailed to the edges, and without crevices.

On one ſide a round opening muſt be cut to receive the mouth of a quart TIN POT from within; and at ſuch a diſtance that the pot may not be nearer than *an inch* from the ſide, and *three inches* above the bottom.

The QUART POT, without a handle, is to be punched round the ſides as full of holes as poſſible, within an inch of the top (except about two inches, which need have but few), as alſo in its bottom. The holes ſhould be as large as thoſe of a flour-dredger. The pot is to be fixed in the circular opening by flat-headed tacks, with the part having the feweſt holes next the bottom.

Another

Another APERTURE is to be cut on the right of that for the pot, six inches in length and four and a half wide, to receive a pane of glafs; it is to have a fhutter to let into a bevel at top, and reft on a ledge at bottom, and to faften with a button. A wooden or cork ftopper muft be fitted to the pot. It will make the box more con-venient for vifion, if a fmall window three or four inches fquare is made in the *back*, about three inches diftance from the bottom.

OBSERVATIONS.

Without a great quantity of holes in the pot, each at leaft one-eighth of an inch in diameter, the matter for fuming will not burn freely, and will thereby hinder the effect defigned. The pot is placed an inch from the fide, that the bees in falling may not lodge or be obftructed in their paffage, and thereby *fcorched.* For a like reafon the pot is three inches above the bottom. The circular form of the pot prevents any confiderable number of bees from being detained thereon The part on one fide having few

6   holes,

holes, is for laying the fuming fubftance on.

Annexed to the FUME-BOX is a *frame*, pl. 1. fig. 5. to nail on its edge. It confifts of a *hollow fquare*, the rim three inches broad, and three quarters of an inch thick ; the infide hollow, to be equal to that of the box ; the other parts to extend over the outfide.

This frame is intended for placing *full boxes as well as hives over it* ; and therefore, to fuit it to that purpofe, its *corners* have four fmall pieces of wood faftened in, to adapt it to the circular bottoms of the hives. By the breadth of the *rim*, it will likewife admit any common-fized hive.

OTHER INSTRUMENTS are, a long thin and broad *knife*, with a fquare end, and a fquare wooden handle:

TWO BRASS PLATES, OR DIVIDERS, and *two flips of double tin*, of the fame length as the plates, and three inches broad ; or in lieu thereof, two old faws without handles, and their teeth taken off.

No. II.

The Material for Fumigation.

In my former treatife I flightly mentioned a method of *ftupefying bees*, but have fince that time heard that fome perfons on trial could not fucceed.

Having always been in a habit of *driving*, I did not give the fubject that attention which it fo juftly deferved. But reflecting on the great advantages it was capable of could the difficulties be furmounted, I ftudioufly applied to experiments, to accomplifh this defirable end, which I now fubmit to my apiarian friends.

The SUBSTANCE beft adapted for this purpofe is the *Lycoperdon*, or great puff ball. It is likewife called frog cheefe, mully puff, punk-fift, and by various other names; but I fhall mention it only under that of PUFFS, in the fequel.

In good foils it frequently grows as large as a child's head; commonly as large as the

double

double fift. There is alfo a fmall fort, about the fize of a fmall apple, but of a very weak quality.

Both forts grow on dry pafture grounds, and in woods; and thrive where mufh-rooms do, and nearly about the fame time, or either fooner or later,—or fometimes not at all.

They are to be gathered in *dry* weather, if poffible, and as foon as full grown, which is in about eight days. They then begin to turn brown and powdery, and are *then* moft fit for the purpofe. But if not come to their growth, when dried, they become too hard to hold fire. Prefently after the puffs are gathered, expofe them as much as pof-fible to the funfhine to dry; or for want of that, in a dry fhade, &c. fecure from wet or dew. Drying them by the *fire* makes them hard, however moderate the heat.

Care muft be taken to preferve them dry, in paper bags, in a dry room, till they are wanted. For as they feldom come in fea-fon early enough for ufe, they fhould be preferved in readinefs for next fummer. They are to be laid on the hearth for an

5   hour

hour or two, the evening before they are to be ufed, to expel from them the dampnefs their fponginefs makes them liable to; which would render them unfit for burning freely. Age likewife has the fame effect, bereaving them in part of their ftupefying power. Puffs found in autumn in woods, or under hedges, being dried by a moderate fire, though not fo good, may do for want of better. Puffs kept longer than the fecond year, retain little virtue.

Thofe puffs which in a dry feafon have become mature, light, and dry, burn the *beft of any*. The lighter and more *fpongy* the puffs are, the readier they burn. Thofe that are gathered in, or foon after, wet weather will be very tardy in burning, being deprived in a great meafure of their virtue, however dried afterwards; as will thofe that have been dried, but fuffered to get wet again, but which redrying will not reftore.

When a *wet* feafon, or any other caufe, has hindered the acquifition of puffs of a good quality, they fhould be fteeped in a *folution of nitre* (falt petre) in water, viz. a
tea

tea fpoonful of nitre grofsly powdered, to a pint of water. After the puffs are foaked therein, they are to be well dried, and, thus treated, will quickly take fire, and retain it.

But if puffs are very bad, rub a piece of camphor, of the bignefs of a pea, to powder, and then add a little linfeed oil. This being fmeared lightly over a puff, will immediately take flame by a candle; blow the flame out, the puff will continue to hold fire, and fume till reduced to tinder.

It is to be noticed, that moft dry puffs will readily hold fire in the *open air :* but when introduced into a clofe box, excluded from frefh air, they CEASE TO FUME. And fhould frefh air be admitted, it would counteract the ftupefying quality of the fume already admitted, and delay the operation, or make it wholly unfuccefsful ; therefore the leaft frefh air poffible fhould be admitted.

No.

No. III.

The Method of Fuming.

TAKE as many pieces of puff, each about the fize of an egg, as the fume pot will hold without preffing; lay the pieces on embers, or live coals, in a chaffing-difh, or the like: when they appear to fume well, put them nimbly into the pot of the fume-box, and immediately ftop the mouth. The hive or box of bees being *previoufly* fet over the box (with all crevices ftopped, that no fmoke may efcape), in about fifteen or twenty minutes the bees will be STUPE-FIED, and fall from their combs into the fume-box. This will the fooner happen if the hive or bee-box is now and then gently tapped on the top. When the fmoke firft arifes, it caufes a great buzzing among the bees, which gradually ceafes as they become fenfelefs; and then they may be heard to

drop

drop down, and will recover again in about the fame fpace of time on the admiffion of frefh air, and without receiving the leaft injury.

A flight fuming will at all times render them very peaceable, though not quite infenfible.

A fmall portion of brand may be laid on bad puffs when they are firft put in, left they fhould not retain the fire. The *ftupefying bees is in no wife prejudicial to them*, fince they foon return to their wonted labour and activity, as if no fuch operation had been done. Nor do they afterwards fhow any refentment upon that account, which is always the cafe after *driving*.

No. IV.

A Method for Cottagers who are not provided with Dividers.

MAKE a HOLE in the ground fomething *lefs* than the circumference of the hive, and eight

eight inches deep; spread a cloth to cover the bottom and sides.

In the evening take a stick seven inches long, having a slit in its end to receive a piece of puff about the size of an egg; light it, stick the other end in a clod of clay, and instantly place a hive of bees over it; and they will become as easily stupefied as when suffocated by brimstone. If one piece of puff is not sufficient, put in two or three upon sticks.

No. V.

OR, instead of a hole, a circular RIDGE OF EARTH, nine inches in height, with the inside hollow, and suitable to support the hive, when set over it. Leave a *part* of the ridge open to put in a fuming-pot, which may be a small earthen pan, an old tin pot, or the like, in which put the lighted puffs, and cover the pot with an old funnel (the pipe off) with many holes in it, to keep the bees from falling on the burning puffs. Immediately

mediately on putting the pot under, ftop the opening by a fod of earth, made ready for that purpofe. If the puffs fhould not hold fire, run a wire, or fmall ftick, through the fod, to let in a *little* frefh air.

Or, an EMPTY HIVE may be ufed for this purpofe, in lieu of the earth; turning the hive upfide down, and fetting another thereon.

Or, the ridge of earth may be made on a board, and fo be more convenient to be removed near the hives.

If the hives are not of equal circumfe-rence, two fticks, of the due length, with two others nailed acrofs them, and laid over the hive, pot, or kettle, will conveniently fuit any hive you have.

Thefe methods are defigned for ftorified hives ONLY; it being of *no ufe* to fave the bees of fingle hives.

No.

No. VI.

The Use of Dividers.

IN SEPARATING STORIFIED HIVES, thruſt in one of the braſs dividers firſt, with its turned end *upwards*, between the two hives; then ſhove in the other with its *turned end downwards*, and ſlide it under the firſt. At the ſame time, an aſſiſtant is to keep both hives from ſlipping out of their places. When the apiator withdraws the upper divider, and hive thereon, the aſſiſtant is firmly to keep the *under* divider from moving with one hand, and with the other keep the under hive ſteady. The apiator, in drawing the hive towards him, muſt move his hands gradually under the divider, till nearly half is withdrawn; he will then feel it upon a poiſe, ſtill keeping the divider cloſe up to the hive, lift it gently and care-fully up, and ſet it on the fume-box, placed by him in readineſs. The aſſiſtant, in the mean while, is to place another empty hive

over

over the ftock in lieu of that taken off, or
a cover, as the cafe may require. Keep the
hand on the còver, or empty hive, and withdraw the divider. If the divider do not
eafily come out, ufe a pair of pincers.

Sometimes the irregularities and fnags of
the broken binding or ftraw of the hives
greatly obftruct the free entrance of the divider. To obviate this, it is proper to have
two SLIPS of double tin, fifteen inches long,
and four wide : they are to be fhoved in on
the right and left fide of the hive ; introducing them at the middle of the fides, and not
at their ends, they will then generally pafs
eafily. But if any impediment occurs, run a
broad knife between the edge of the hive
and the tin, and raife it a fmall degree at the
point of obftruction. Or, if it arifes from the
under hive, the knife is to enter *under* the
tin, to difengage it. The flips having paffed
nearly to their whole *width*, the dividers are
to be fhoved in at the *back* or *front* of the
hive, as fhall be moft eafy, and *under* the
flips ; by which means they will enter with
great facility. Obferve to turn their ends as
before mentioned. If the dividers enter at

Q   the

the back, a perfon muft hold his hand againft the door-way, to prevent the ftopping from being fhoved out.

Particular care fhould be had, in taking out the dividers, to fet them upright againft fome fupport, or to lay them flat, to prevent their being *bent*, which would render them unfit to keep the bees clofe in.

No. VII.

To Storify.

To *fet on* a DUPLET, loofen the cover of the ftock, and flide a divider underneath it, keeping one hand on the cover. Take it off as foon as the flider is adjufted; then fet an empty hive upon the divider, and keep the hive faft while it is withdrawn. Early in the morning, or in the evening, will be the propereft time to do it: a pair of gloves only will be needful.

To place a DUPLET UNDER a ftock, fet a ftool behind the ftock; fhove the divider

under

under it, then lift the hive and flider on the
ftool; fet an empty hive (with its cover off)
and floor in the place of the ftock, which
lift thereon; pull out the flider with one
hand, while the empty hive is kept fteady
with the other.

A TRIPLET is to be managed in the fame
manner.

No. VIII.

Deprivation, or Separation of Hives.

FIRST, a TRIPLET is to be taken in the
evening. The dividers are to be introduced,
as by No. VI.; the feparated hive is to be
placed on a floor, at fome diftance, and then
the door unftopped. In about an hour after,
or the next morning, if the bees in the triplet
are quiet, as alfo thofe of the ftock, there
are queens in both; but if not, fhut the door
of that taken, and fet it over the fume box,
and proceed to fume, as by No. III.

If

If an under or nadir hive is to be taken
from a *double* or *triple* hive ftock, the
fame method is to be ufed ;—only the two
upper hives are to be taken off together, and
placed on a ftool till the nadir is taken away,
and then fet on a frefh floor in its old fituation.

Second Method of taking Triplets

Is, for a ftout man to lift up the triplet,
ftock, floor and all, and take them to fome
apartment, in which a ftrong form or bench
is firmly prepared clofe to the wall, and
to place them on that. It may be done any
time in the morning, if the bees are very
early fecured from coming out. Follow the
directions of No. VI.; only the operation may
be more fecurely done, without being in-
commoded by the bees of the apiary, when
at a diftance from them; and being againft
a wall, the hives are kept more fteady during the infertion of the dividers. If the
middle hive feems full of combs, and has
not much brood, that alfo may be taken.

On the GENERAL DEPRIVATION, the
hives

hives taken off muſt be ſet apart in another part of the garden, to diſcover which
have queens; as alſo of the ſtocks. And if
any are without, the hive taken from *it*
muſt be reſtored, and remain ſome weeks
longer. The further fumigation is to be
deferred till the next day after taking. It
is to be noted, when hives are *lifted on the
fume-box*, it ſhould be *on the divider*, which
is then withdrawn, by which means no bee
can eſcape. The ſtupefied bees are always
to be put in an empty hive, and placed before the ſtock, on ſome ſupport.

To SEPARATE DUPLETS, is ſo obvious
from what has been written, as to preclude
further directions.

COTTAGERS muſt purſue the methods of
No. IV.

Many times the EDGES of STRAW HIVES
will be ſo uneven as to ſuffer the bees to paſs
under them, ſo as to be very troubleſome on
the introduction of the dividers. To remedy this default, prepare a narrow ſlip of
coarſe linen cloth, about three inches wide,
and of a length ſomewhat more than the

Q 3 circum-

circumference of the ftraw hives in ufe.
Two fmall wire hooks are to be fixed at one
end.   This cloth is to be thoroughly wet,
and drawn round the body of the hive, about
an inch and a half above the bottom edge.
When the dividers are to be ufed, raife the
edge of the cloth, juft high enough to fuffer
them to pafs a little under, and let the cloth
drop clofe round.   Its weight will render it
fo clofe as to exclude any bee from paffing.

It will many times happen, that a few
bees will ftill remain in the hive, notwithftanding the moft powerful fumigation, by
having fecured themfelves in the empty
cells; or by the fume not being ftrong
enough when *firft* put in.   In fuch a
cafe, throw a cloth over the hive, and take
it into a dark room, there to remain till the
next day; when gently drumming or tapping on the fides and bottom of the hive,
they will rife to the edge of the combs, and
fly home, without fhewing any anger.

When a hive is cleared of bees, the brood
combs fhould be properly placed in an empty
hive, *inverted on a divider*, and fo placed
over

over the ftock.   This muft be done very leifurely, left it provoke the bees to deftroy the young.   If one hive will not hold them, put the refidue in another, which fet over fome other ftock.

Boxes are much eafier feparated than hives, from having their edges more even, though the like obftacles will fometimes happen; and which are relieved by the fame means as for hives.   But the ufe of tin flips will not be requifite.

It will be advifable for the unexperienced to practife the manner of operation by trials on empty hives with a weight laid over them, before they attempt with hives of bees.

Care muft be taken, that as few bees may be killed as poffible; efpecially where the queen's death would be the ruin of the hive.

COTTAGERS fhould feparate the combs from the bottom hive the night before, by a knife; when they are to take the hive off, give it a kind of twift, and then lift it on the ridge of earth, as by No. IV. while a perfon *inftantly* throws a cloth over the top of the hive left, there to remain till next morning;

Q 4   then

then placing the edge of the ftraw cover juft
under the cloth, fhove it nimbly and clofely
with the right hand, while the cloth is kept
fmooth with the left ; by which means the
bees, and pieces of combs, that were lodged
on the top, will be pufhed off by the fliding
in of the cover.   But if the cover does not
fit clofe, ftop the chafms with mofs till cold
weather comes, when the obftruɗions may
be pared away with a knife.

No. IX.

*The Re-union of Swarms with their Stocks, or
with each other.*

HAVING hived a recent fwarm, take it
to a diftance from the apiary, lay a cloth on
the ground, and ftrike the edge of the hive
thereon ; the bees will fall out in a lump.
With a fpoon tenderly divide them into three
or four parcels, putting them into as many
pans, fieves, &c. and fet each parcel at a

4   con-

confiderable diftance from each others' fight. Thofe parcels which have no queen will foon return home again. That which remains take to a darkened room, and fume, as by No. III. This done, turn them out upon a table, and with a fmall ftick difengage a few at a time from each other, and look attentively for the queen.' If not found in the firft number, ftrike them off the table into an empty hive, and thus proceed with the reft. When fhe is found, inftantly feize her between the finger and thumb, and put her into a phial with a notched cork, and about a dozen workers with her, to keep her warm and eafy. *Infpect* the remainder of the parcel, left there fhould be another young queen. Include them all in one hive, and fet them down before the ftock, to which they will gladly unite.

 But fhould a queen not be found, it is poffible fhe may have fallen down, and been crufhed. In that cafe the bees will foon fhew their inquietude, and return home. If not, give them a flight fuming, and proceed as before, but with more circumfpection.

Keep

Keep the captive queen two or three days, when, if there fhould be no occafion for preferving her, death muft be her portion. For, if let loofe, fhe will return to the ftock, and occafion a repetition of the procefs. Or elfe make an artificial fwarm with her, if wanted.

By the like means, AS MANY SWARMS as rife may be added to the ftock, or united with *one another*, to form a powerful ftock of themfelves. Only *then* keep the bees in the hive, with a cloth over it, and take them out by a fpoonful at a time, to examine them, pulling the cloth over after every fpoonful, to prevent their reviving too foon.

OR, SWARMS may be united, three or four, or more, together, to form a ftock, as directed at pages 99 and 100, or at deprivation time.

When two queens rife together with a fwarm, and are hived, but prove hoftile to each other, fumigation will reconcile them. The firft queen that recovers will be acknowledged, the other flain.

If a fwarm that is to be united is tumultuous and mifchievous, the giving them a
                                    flight

flight fuming will make them more tractable. It is worth remarking, that bees are often adverfe to receive ftrangers at one time, but will cordially receive them at another; therefore they muft be humoured.

No. X.

Captivating the Queen of a Stock.

FUME the ftock, and examine the bees, as in the foregoing article. Sometimes *fhe*, as well as fome other bees, will evade the effect of the fmoke, by entering the empty cells (which is equally the cafe even with brimftone), and therefore muft be proceeded againft as before directed. To diftinguifh a queen, a previous knowledge fhould be acquired, by infpecting the bees that have been fuffocated. A queen may be attached to any part, by paffing a filk thread round her neck, and clipping off part of one wing. Where fhe is fixed, the fwarm will furround and never quit her. Or a queen may be

capti-

captivated thus : Put the bees that have a
queen into a hive or box, whose top has
long slits of *only five thirty-seconds* of an inch
in width. The working bees, by much
tapping on the sides of the box, or by blowing the smoke of tobacco in, will issue out,
and leave the queen behind, as she will not
be able to pass the slits, if accurately made.

No. XI.

Outliers to recruit weak Stocks.

AT the close of the evening, place a floor
on a level with, and to touch that of the out-
liers ; bring the *weak* stock pretty near ;
then with a small stick very leisurely stroke
the out-liers down on a vacant floor. Instantly take away the stock, and set it at a
little distance, while an assistant places the
weak stock over the floor of out-liers, its
edge being kept raised by a wedge. Let
them remain till day-break, by which time
the idlers, in all probability, will have afcended;

cended; when, taking away the wedge, re-
place the ſtock in its former ſituation, and
the other at a conſiderable diſtance.

But when a great quantity of bees cluſter
round the body of a hive, an empty hive
ſhould be placed near; when lifting the ſtock
upon the empty hive, idlers and all thereon,
they will ſoon find and embrace the new
accommodation.

Another method is, to ſpread a cloth underneath, and by a bruſh or watering pot
ſprinkle water over them; by which means
they will be unable to riſe, and may be
bruſhed off on the cloth, and put on the
floor of an empty hive, and the weak ſtock
over them.

No. XII.

To unite a queenleſs Stock to another.

WHEN a ſtock in *ſummer* has loſt its
queen, ſtop the door immediately, till the
other ſtocks have done work, *then* open

it for about an hour, and then ſtop it again; ſlide under it the divider, fume it, put the bees in an empty hive, and ſet them over another ſtock. By this means, as they gradually acquire vigour, they will aſſimilate with the ſtock, without any diſturbance. The hive of combs taken moſt likely will have much brood therein, which is to be diſpoſed of as before mentioned, and what honey there is is at the owner's ſervice.

No. XIII.

To unite weak Stocks or Swarms in Autumn.

IF, through inadvertence, weak ſtocks or ſwarms have been retained till autumn, and one of them has a ſufficient winter's ſtore, incorporate the lighteſt with the ſtrongeſt, by fuming each ſeparate, and then placing the weak one over the other. When the bees recover, they will unite without ſtrife, and the ſupernumerary queen be caſt out. If they are both poor in honey, but ſtrong

in

in numbers, they will form a good ſtock, if a good hive of honey is placed over them. Otherwiſe, ſuffocate them, and take the honey, and ſave the brood, if any.

Cottagers may unite them by turning one hive bottom upwards, in a cold day, for ſeveral hours, till the bees become chilled and feeble: the combs are then to be taken out ſeverally, and the bees bruſhed off upon a table, and the queen taken from them. Then put the bees into a pail, pan, &c. lay two ſticks acroſs, and place the other hive over it; cloſe the joining with a cloth, all but the door-way. Let them ſtand thus two or three days, in which time they will have united. If afterwards the hive ſhould be found too light, the bees ſhould be fed.

After all, this is but a ſhift, which ſeldom anſwers. Had they been incorporated in ſummer, they would have turned to good account.

No.

No. XIV.

Driving.

Pass a divider under the hive to be drove, and then tie a cord acrofs it and the divider; turn the hive upfide down on an empty hive, bucket, or fomething convenient. Place the fume-box, reverfed, over the hive, (firft taking the cord off), and gently withdraw the divider, taking care that the door of the hive is well clofed ; then with two fticks beat, as though drumming, on the fides of the hive (all but that fide next you) and at the bottom, not very hard, but very quick, ceafing a little at intervals. In about fifteen minutes the bees will begin to be terrified : hearken whether they make a great buz, and whether a buz is likewife in the box, for by that it may be guefled that many are afcended. Some one fhould hold the box fteady while the drumming is made, or it will fhake, and let the bees out. The box may then be fafely lifted up on its fide oppofite to the light (for the room

fhould

fhould be almoft dark), and the bees will fly
directly towards the light. Hold the box
fteadily between your fide and arm, and
with the other hand continue tapping round
the fides. The bees by this become tame,
and will gradually crawl up from the hive
into the box, with loud buzzing; and the
more fo when the queen afcends, for then
the reft will *foon* follow; but till that happens they rife with great reluctance.

By chance a few may be left behind,
which may be drummed out the next day.

If no fume-box or divider is provided, a
common ftraw hive may be ufed inftead;
and the ftock lifted on it (when inverted) over
night. In the morning, ftopping all chafms
and the door, tie the two hives faft with a
cord, and invert them, and then proceed as
above.

The driving of bees renders them very
peaceable and tractable, fo that they may be
fafely taken up in the hand; though not fo
completely as by fumigation. They may be
turned on a table, feverally divided, infpect-
ed, and *the queen taken from them.* But this
peaceable difpofition continues but a little

while in either method; fo that the operator muft be as expeditious as poffible. Except,
when the bees are kept in an empty hive
two or three days, it will make them extra-
ordinarily tame.

DRIVING WILL BE USEFUL as a fucce-
daneum for puffs, in feafons or circumftances
when they cannot be had.

To NATURALISTS it may be of advantage, by enabling them to inveftigate the
properties of thefe wonderful but irritable
infects, while in a ftate of vigour, more fa-
tisfactorily than by *fumigation* or *immerfion.*

No. XV.

Show-Box for Amateurs.

THIS BOX, or FRAME, is to be made of
rattan or mahogany, without top or bottom.
It is to be nine inches high, and two and a
quarter wide, clear in the infide, *exactly*; and
twelve long. There is to be a pane of clear
glafs

glafs on each fide, as large as the frame will admit. The glaffes are *not* to be let into a rabbet, as ufual, but to flide up from the bottom to the top within fide, under four fmall tenter hooks, and ftopped at bottom by a fmall fcrew, fo that the glaffes may be taken out occafionally. There are to be two half-inch *fhutters* on the outfide, to faften in a bevel (not to flide) at bottom, and with a button at top.

The top is to have a bar one inch wide, and the full length of the frame, and is to be let in at each end fo as to be flufh with the top, and at half an inch diftance from each fide of the box.

A door-way is to be cut at one end, one inch and a half long, and half an inch high : this is to be efteemed the front. At the other end or back, a like door-way is to be cut ; and another three inches higher.

On the top edges of the box are to be two narrow flips or ledges faftened ; between which are to lie (not to flide in a rabbet) two pieces of glafs, each half the length, and fufficient in width to cover the top between

the

the ledges. A wooden loose cover muft clofe
the whole.

It muft have a *loofe* FLOOR two inches
wider than the box; and little abutments
fhould be added at the corners near the bottom, to give the box a fteadinefs fufficient
to prevent its being turned afide.

OBSERVATIONS.

THE narrownefs of the box, and its having but one bar, is intended to prevent the
bees from making more than QNE COMB,
which they would do if it was a *quarter* of
an inch wider.

By being confined to *one* comb, the mi-
nutiæ of their tranfactions are expofed to
view on both fides; the queen's not excepted: a difcovery I fufpect not to be fo fully
obtained by any other means known to the
public.

A window, full fouth, is the propereft to
place the box in. A fituation the leaft expofed to wind is neceffary for their fucceeding. The bees are ufed to great heat, and
commu-

communicate much themfelves, which caufes a great indraught of cold air to be very prejudicial and difcouraging.

I invented this box in the year 1783, when removing to this fituation (which is an extraordinary windy one, beneficial to my own health, but not propitious to bees); and not having a convenient afpect to fix the box in, the wind greatly impeded their labour, and fruftrated my defigns; except one year, which being tolerably favourable, my purpofe was in *part* anfwered.

The two door-ways in the back of the box are the readier to introduce troughs of food, in cafe the bees, through bad weather, are hindered from collecting, after being firft put in, or at any other time. The two back door-ways are to be always clofe ftopped when not ufed.

The reafon why the panes of glafs are not to be fixed in, is, that in cafe of accidents they may eafily be repaired.

No. XVI.

Management of the Show Box.

PROCURE a flip of deal, of the length of
the box, one inch and a quarter broad, three
quarters of an inch thick : pierce fmall holes
in it, at equal diftances, four on a fide, into
which put eight flight fticks, four or five
inches long, and thus form a ftage, cutting
off all irregularities at the bottom.   Place in
it a *thick empty virgin comb*, four or five
inches in length and breadth.   Introduce it
as far up the middle of the box as to touch
the bar; faften it at the ends by two fine
and long fcrews, paffed in at the front and
back of the box.   *Or*, the ftage may be
hung to the bar by four ftrings (horfe-hair
will be beft) over the bar let into grooves, and
tied on the fide of the bar, that there may
be no obftacles above the level of the box.

Having procured a QUEEN from a *fwarm*,
cut her wings half off, put her into the box
at the top, the door being ftopped ; then put
a PINT of *fumed* bees, including ten or
twelve.

twelve drones, into the box with her. A
lefs number of bees than a *pint* will be too
few ; and a greater will fo much crowd the
comb as to prevent the view defigned.
Clofe the top by one half of the glafs, and
the other by a perforated piece of tin. Whenever the door is unftopped, both pieces of
glafs muft be laid over, or there will be too
great a draught of air. Throw over them
a cloth, and let them remain till the morning ; then unftop the door fo as to admit a
paffage of two bees at a time. If on the
fecond day the bees feem contented, entirely
unftop the door, and give them a trough of
food. Refrain from opening the fhutters
for four or five days, and then but feldom,
till they have begun to collect, and repair
the combs, or it will difguft them, and caufe
a defertion of the box, which will fometimes happen notwithftanding.

For the queen and her fubjects, being ufed
to a much greater heat, to a larger fociety,
and a more commodious habitation, will be
very much difpleafed at fuch a fcanty tene-
ment, and not foon reconciled to it.

But however difgufting it may be, if the

queen *does crawl* out, or her fubjects fwarm
out, the one muft drop, and the others,
though cluftered on fomething near, muft
return, and the queen may be found under
the window, and again returned into the
box. The clufter, being fecured, is to be
introduced to her.

If great winds annoy them very much,
they will emigrate, though they have en-
riched the box with honey and brood. The
clufter that fettles may be fhook into an
empty hive, and fhook out again upon a
table, and the box placed near them, when
they will foon join the queen.

For the purpofe of excluding the wind,
it is advifable to have a tin trough, of the
fhape of a T ; the long end to fit the door-
way of the box, and to be open at the other
end, as well as at the ends of the crofs tube.
Corks are to be fitted to them, that either of
them may be ftopped in the point from
which the wind blows.

When the weather proves cool and chilly,
cover the box with a woollen cloth.

When the bees are wanted to relinquifh
the box, flide a divider under it, and fet it

over

over the fume-box ; shove the box as near
the edge of the hollow, as its width, and
withdraw the slider the like width, and the
bees will have a free opening to fall into
the fume-box. Fume them according to
art.

The box must be set on a board in the
window, and so that no bee may have
egress to the room ; obferving the like precautions as before advised for window boxes.

POST-

POSTSCRIPT.

J U S T as my manufcript was ready for the prefs, I became acquainted with a Treatife, recently publifhed by Mr. *James Bonner*, of Edinburgh, purporting to be " *A New* " *Plan for fpeedily increafing the Number of* " *Bee-Hives in Scotland,*" &c. Upon a careful perufal, feveral paffages in that work feemed worthy of notice; but not to alter the body of my own, I here give them feparate, with a few brief remarks.

Mr. *Bonner* is a ftickler for the *Schirachean* doctrine of raifing young *queen bees* at pleafure, in order to form *artificial fwarms;* and oppofes thofe of a contrary opinion, though fortified by numerous experiments of refpectable naturalifts, at home and in Germany, feveral years after thofe of Schirach.

2   The

The fubject of difpute is of little confequence, as not being advantageous for the *general* ufe of thofe who feek the beft method of producing the moft honey and wax; nor is it eligible for the purpofe it was defigned for, viz. Artificial fwarming.

The champions of both fides exprefs their doubts of its general benefit. B. himfelf, in particular, fays, " It is not a great number of hives that will produce the greateft " quantity of honey and wax, but only real " good ones. I alfo doubt whether more " hives can be reared by *this method*, as our " bees generally produce more queens natu- " rally, than they are able to fupply with a " fufficient number of common bees to " compofe a fwarm with; as appears from " their killing the fupernumerary ones;" and therefore he " prefers natural fwarms."

Schirach's method is by a double hive, and the bees are compelled to afcend into the upper one by the fmoke of rags, &c. A piece of brood comb is cut out, of four or five inches diameter, containing a maggot or maggots, precifely of three days old, and properly placed in an empty hive, together

ther

ther with part of a comb of farina, and
another of honey: about a quart of bees
is then to be introduced, and the hive
ftopped up, except a fmall paffage for air,
and fo remain three days. There will be a
great tumult and noife in the hive for fome
hours, when it will fubfide, and the bees will
begin to build a royal cell. The fourth day
an opening is to be made of a quarter of an
inch, that the bees may come out leifurely.
After roving about for fome time, they will
return to their hive. It fhould be done in
the fpring.

B.'s procefs is, I think, an improvement:
he *drives* the bees out, then cuts a piece of
comb out that has feveral maggots, of various ages, and placing, &c. and then fets the
hive at a very confiderable diftance from the
apiary, *without ftopping the bees in.* This I
underftand to be B.'s method, for he feems
referved as to an explicit explanation. I
make no doubt but the ufe of the *puffs* will
be found preferable.

Shirach's ftopping the bees was ill judged,
and what, perhaps, occafioned my bad fuccefs.

It

It is fomewhat ftrange that Mr. B. fhould have purfued his refearches, without the advantages of bee-glaffes, or bee-boxes, but confined himfelf to *ftraw hives* of the common form holding two pecks and a half, and occafionally eeks.

His principal *dependance* for *rearing a great number of ftocks*, is by providing a *fufficiency* of pafturage adequate thereto; but the waxen caftle he has raifed for this purpofe feems to have been built on a hill of fand.

He fuppofes a perfon to begin with five ftocks, which the fecond year will be increafed to ten, and fo continue to increafe in a duplicate ratio for ten years, which will then amount to 2,500. He fuppofes likewife, that if each parifh of Scotland had twenty hives in May, the amount of the eight hundred parifhes would be 16,000. Suppofing each of thefe hives to throw out one fwarm in September, we fhould have 32,000. On thefe principles, with proper management and tolerable feafons, in the fpace of feven years the ftocks would increafe from 32,000 to 2,048,000; and after his draw-backs, his loweft eftimate is a clear million,

million, producing 4,000,000 pints of ho‑
ney, and 1,000,000 pounds of wax.

On the suppofition that bees will increafe
double every year, and therefore that five
hives the firft year may increafe to ten the
fecond year, &c. I will not difpute ; but
will there be *double the quantity of honey and
wax?* I doubt, not : for, fuppofing the five
hives (the bees of them) can only collect
from the vicinage, as far as their flight for
pafturage ufually extends, enough to fill
their five hives ; the fecond year being
increafed to *ten,* the fame quantity of flowers
will only yield the fame quantity of honey,
admitting the feafon fimilar to the firft. I
infer, therefore, that the produce will be no
more, though double the number of bees.
To this we may add (which B. acknow‑
ledges) that feafons are often bad ; rendering
hives impoverifhed inftead of increafing, and
that they often die in the winter. The fecond
link of this golden chain being broke, down
falls the whole mafs of honey and wax ap‑
pending thereto, and there I leave it.

No! fay its advocates, that is not fair!
We can increafe the flowers in proportion

to

to the number of bees. Can the cottagers extend their land? or will they extirpate from their little allotment the vegetables of their daily support, to give place for bee-flowers?, Will gentlemen (whom B. chiefly addresses) plough up their grass and corn lands, to cultivate such flowers? Surely corn and cattle are of more value than honey! We had better be without honey than bread. But B. has a resource in heath, which covers, he says, *more than half of Britain!* If true, I am sorry to hear it; and hope most part of it will speedily be ploughed up for corn, though it should prove the ruin of this *new plan of increasing of bees.* I should sooner prefer Virgil's method of raising bees from a dead heifer, or of Sampson's procuring honey from a dead lion.

I sincerely hope, as Mr. B. has been a practitioner for twenty-six years, he has accumulated a snug fortune, to compensate for his labours and ingenious discoveries. But as his native land so much abounds in white clover, heath, furze, &c. it is wonderful that honey sells at ten-pence and twelve-pence

pence per pound, at Edinburgh. It is alfo
obfervable, that he gives no account of the
produce of his own apiary, and only five
inftances of other perfons', of whom he
bought honey and wax. To *one*, in particular, he paid five pounds for ONE HIVE,
which was weighed in the market-houfe
of Edinburgh; but unluckily he omits the
weight or dimenfions of the hive. The read-
er, therefore, is left to his own calculations.

Mr. B. befides his grand refource of flowers, relies on *preferving the bees* of the ftocks
taken, and uniting them with the ftocks
left.

I think his ingenious method of fwarming deferves a place here; and I recommend it to a trial, as it will be too late for
me to do it. My work, I hope, will be
printed before the feafon arrives; and my
age, and increafing infirmities, forbid a
longer delay.

" Suppofe one drive all the bees out of a
" hive, and thereby make an *artificial fwarm.*
" If the old hive has a royal cell in it, by
" introducing into it about five thoufand
" bees, they will hatch out the young queen,

S " with

" with all the eggs and nymphs in the cells,
" and render it a flourishing hive. The
" method of introducing the common bees
" is as follows : Let a strong out-lying hive
" be removed from its usual situation, about
" ten A. M. and place the hive that has no
" bees on the spot where it stood ; the bees,
" on their return from the fields, will enter
" it, and finding plenty of honey, and abun-
" dance of eggs, will rear up the young bees
" with great alacrity." But here it may
be asked, Suppose there happens to be *no*
royal cell in the old hive, how are we to
proceed ? On my plan, instead of *driving*, I
would *fumigate* them out; then inspect whether there is a royal cell ; and, if not, return
the bees into the hive. But if there is a
royal cell, cover the hive of fumed bees with
a cloth, and let an assistant take it to some
distance. In the mean time, carefully examine the old hive, to be assured that the
queen is not left behind among the combs,
as she is frequently one of the last that falls.
Being satisfied on this point, place it on its
original stand. The bees, on their return, &c.
—The hive with the fumed bees should be
confined

confined till night, to be certain that the queen is with them; for, if not, they will soon shew it by their uproar, and, in consequence, must be taken before the stock, and set bottom upwards. But if they remain quiet till night, take them to a very considerable distance, in another garden or field. An empty hive should be set in lieu of the combed hive, during the operation, to amuse the bees as they return from the fields.

As Mr. B. approves the Shirachean doctrine of a common egg being capable of becoming a queen by the nursing of the workers, why should he insist on there being a royal cell in the hive? when common eggs would serve the purpose; only causing a delay of a few days before a young queen, so reared, would be capable of laying eggs.

Another method he gives of artificial swarming is, " to take all the bees out of " the hive, and put into it a considerable " number of common bees, who will hatch " out the brood, and rear them, and often " succeed very well. But this plan is liable " to some imperfections; for, from the time

" the

" the old queen is taken away till the young
" one is fit to lay eggs, will be twenty-five
" days; during which space not a single
" egg can be laid. To which add eighteen
" days more, before the eggs can be of any
" service. It is evident that the best part
" of the honey season will be over, and
" consequently, by autumn, the hive cannot
" be replenished with bees. If I intend to
" kill a hive of bees in autumn, it seems
" best to take away the queen at the end of
" July, leaving a great number of bees in
" the hive, which, having but few bees to
" nurse up, would collect a greater quantity
" of honey in that period, than if they had
" a queen to lay eggs."——

" In the spring, having two hives that
" had but few bees in each, I put the bees
" of one hive into the other, suspecting, as
" they had both bred slowly, there might
" be a defect in one of the queens; and
" hoping that, by putting them together,
" the least healthy would have been killed;
" but the workers of both hives kindly
" united. On turning up the hives twenty
" minutes after, I perceived a few bees clus-

" tered

" tered together. On a clofe infpection, I
" obferved the two queens ftruggling toge-
" ther with the utmoft fury. Being afraid
" of the ruin of both, I feparated them, and
" kept them afunder, though they ran with
" great fury along the table in fearch of
" each other. I then took the one that
" appeared the boldeft, and put her again
" into the hive, where fhe was kindly re-
" ceived. When a duel takes place between
" two queen-bees, the workers commonly
" kill one of the queens themfelves."———

" In November, December, and January,
' bees eat very little food, as any perfon
" may be convinced by weighing their
" hives in the beginning and end of thefe
" months. But if he will weigh a hive in
" the beginning of March, and likewife at
" the end, he will find a confiderable de-
" creafe; for the bees, having now much
" exercife, eat more honey in that month
" than during the three cold ones, and
" three times as much in May as in March,
" owing to the increafe of brood,

" In a mild winter they eat more
" than in a cold one, which enables them to

" *hatch*

" *hatch earlier*, and increase the number of
" bees in the hive. In a very cold winter
" many stocks die; whereas, in a mild one,
" very few. In the midst of a severe frost
" I have often seen my hives with young
" brood in them : they are, therefore, not
" inactive, but *breed* even before they carry
" in loads."——

  " About Lammas, those who live where
" bee vegetation is early over, especially if
" they keep numerous hives, ought to re-
" move them to the neighbourhood of heath
" grounds, if they should be even six or
" eight miles distance ; and allow them to
" continue in that situation till the heath is
" out of bloom. The risk is, if the wea-
" ther turn out bad in August, the trouble
" will be lost." [Is there no risk of robbers?]
" When bees are placed in a new situation,
" they should not be permitted to come out
" of their hive for the first time in cold wea-
" ther; but kept close prisoners for a day or
" two, or many will be chilled to death in
" searching for their new settlements."——

  " Very little ground will keep many bees
" abundantly at work. One acre of land
          " would

" would not be overſtocked with twenty
" hives, and, confequently, the twentieth
" part of an acre would keep *one*!"—[This
ſtatement ſeems vague and unſatisfac-
tory.]

" *Swarms* ſhould be covered with a cloth
" till the heat of the day is abated, left they
" ſhould be urged to rife.

" Nor ſhould it be omitted to keep a watch
" over them, as they fometimes rife after
" being two or three hours in the hive, and
" though they had begun to work—perhaps
" to ſettle in another place they had previ-
" ouſly prepared. Sometimes, though ſel-
" dom, a ſwarm will fly off, notwithſtanding
" every method that can be uſed to prevent
" it. This happens only in very calm wea-
" ther, when bees have had liberty, fome
" days before ſwarming, to roam about in
" ſearch of a habitation to their liking;
" which if once they find, it is difficult, and
" often impoſſible, to prevent them from
" emigrating to it."——

" If the rays of the fun have been inter-
" cepted by a cloud, or ſhower of rain, in
" the time of ſwarming, the ſwarms will

" pro-

" probably be fmall, as preventing the
" greater part from iffuing. In which cafe,
" let the fwarm be placed where the mother
" hive ftood, for about a quarter of an hour;
" in which time the bees that are returning
" from the fields, will foon make the fwarm
" large enough; and then the fwarm fhould
" be removed to a mile diftance, to pre-
" vent the bees from going to the old ftock.
" When bees are feparated from their mo-
" ther hive by driving, or when the hive is
" fhifted from where it formerly ftood, they
" are infenfible of the change, and always
" fly back to their former ftation; for which
" reafon, every artificial fwarm, or rein-
" forced hive, is to be fet at a confiderable
" diftance."—[Would not removing them to
a dark room, and confining them a day or
two, produce the like effect?]

" A fwarm that efcapes from the apiary
" to a habitation they have previoufly
" chofen, ufually fly to it in a direct line.
" The bee-herd fhould run or ride within
" fight of them, as faft as he can; and if
" obftructions hinder him, he fhould atten-
" tively notice the point of the line, and
" keep

" keep or recover it, to march therein
" ſtraight forward, regarding the buſhes and
" hedges as he goes, leſt they ſhould be ſet-
" tled thereon. But otherwiſe the line will
" probably lead him to ſome apiary, where
" he may claim his ſwarm. I know for cer-
" tain, that a ſwarm will not fly a mile to an
" empty hive ; whereas they will fly four
" miles to take poſſeſſion of an *old one* with
" combs in it."

It is proper here to remark, that Mr. B.
repreſents the ſetting of an old hive of combs
in a perſon's own garden, or apiary, as a
fraudulent practice ; as ſuch hives may allure his neighbour's ſwarms to ſettle therein.
So may a field of good paſture allure his
neighbour's cattle or ſheep to feed thereon.
What, then, muſt he not have better paſtur-
age than thoſe in his vicinity ? If ſtrange bees
viſit his hive, which he ſet, *bona fide*, to
entice his own ſwarms, ſhould any eſcape
unperceived, and his neighbour's bees take
poſſeſſion of it,. *without* being followed by
a perſon who ſaw them riſe, he ſeems to
have a *good title* to keep them ; for who
can ſwear *whoſe* property they were ? They
ſhould

fhould have been better watched. The lofs.
they deferve for their negligence, which I
hope will make all bee-owners more careful in this point, if for no other reafon. No.
honeft perfon will refufe the reftoration, if.
they can make good their claim. If a.
perfon fets fuch hives with a view of tre-
panning his neighbour's fwarms, it is certainly wicked. The *motive* conftitutes the
crime.

. " Driving of bees, to make artificial
" fwarms," Mr. B. obferves, " is very pro-
" fitable, when properly performed by fkil-
" ful *bee-mafters*; yet it always has been,
" and ever will be, deftructive to bees, if
" performed by unfkilful perfons. And,
" indeed, all new beginners may be almoft
" certain of ruining fome hives in their at-
" tempts."

·· T. Wildman corroborates the affertion,
by faying, " It is an *art* not fpeedily attain-
" ed ; yet, till it is, the deftruction of many
" hives muft be the confequence, as every
" one will find, on their firft attempts to
" perform it." To which truth, J. K. fets
his feal!

Mr.

Mr. Bonner, it feems, has been a bee-manager from his youth; and is now a profeffor of the art, and proffers his fer-vice to the gentry of his country, who may be defirous of his affiftance. He appears to be a fuccefsful pupil of the elder Wildman, and like him enumerates feveral *manœuvres* that he can perform, &c. but he does not, like Wildman, divulge the fecret of HOW, which he referves for his own ufe. However, we may fhrewdly guefs, that it is by means of the bee-drefs, by driving, and by the ma-nagement of the queen-bee; by which, to my thinking, any intelligent perfon, con-verfant in practice, may eafily do the like, if any one would compenfate him for his time and trouble of amufing them, which is the only ufe thefe feats feem adapt-ed for.

In a few words—Notwithftanding Mr. B. confidently affures his readers, that his plan is " no *chimera*, or *Will o' the wifp*," many of them, perhaps, may require more folid proofs on which to eftablifh fuch an idea. The more wonderful any thing of-fered for our belief is, the ftronger fhould be

the

4

the evidence. It ſeems requiſite they ſhould know what number of ſtocks B. as well as ſome of his principal pupils, keep: the quantity of ground ſown with bee vegetables on purpoſe: what the quantity of wild bee flowers is in the circuit of their flight: and what the *produce* is, on an average, for ſeveral years, &c. Till this is done, *thoſe that have little faith, but much reaſon, will ſtill doubt, if not diſbelieve.*

I N D E X.

INDEX.

Swarms